Lecture Notes in Chemistry

Edited by G. Berthier, M. J. S. Dewar, H. Fischer
K. Fukui, H. Hartmann, H. H. Jaffé, J. Jortner
W. Kutzelnigg, K. Ruedenberg, E. Scrocco, W. Zeil

13

Giuseppe Del Re
Gaston Berthier
Josiane Serre

Electronic States of Molecules and Atom Clusters

Foundations and Prospects of Semiempirical Methods

Springer-Verlag
Berlin Heidelberg GmbH 1980

Authors

Giuseppe Del Re
Cattedra di Chimica Teorica
Università di Napoli
Via Mezzocannone 4
I-80134 Napoli

and

Via B.Bompiani 15a
I-00147 Roma

Josiane Serre
Laboratoire de Chimie
Ecole Normale Supérieure
de Jeunes Filles
1, rue Maurice Arnoux
F-82120 Montrouge

Gaston Berthier
Université de Paris
Institut de Biologie
Physico-Chimique
Fondation Edmond de Rothschild
13, rue Pierre et Marie Curie
F-75005 Paris

ISBN 978-3-540-09738-9 ISBN 978-3-642-93134-5 (eBook)
DOI 10.1007/978-3-642-93134-5

Library of Congress Cataloging in Publication Data. Del Re, Giuseppe, 1932-
Electronic states of molecules and atom clusters. (Lecture notes in chemistry
Bibliography: p. Includes index. 1. Molecular theory. I. Berthier, Gaston, 1923 .
II. Serre, Josiane, 1922- III. Title.
QD461.D36. 541.2'2. 79-28726

Originally published by Springer-Verlag Berlin Heidelberg New York in 1980
Softcover reprint of the hardcover 1st edition 1980

The Scope of the Review

Recent advances of the quantum theory of molecules have been marked by more or less open methodological controversies. There have been emotional overtones that have actually hindered fruitful research and useful criticisms. The debate is now abating and everybody is realizing that *bona fide* methods, whether containing empirical parameters or not, can all contribute to our understanding of facts. There exists no miraculous or best method; both semiempirical and ab-initio methods do not provide more than approximations, and the important question is the extent to which they can help in correlating and predicting the properties and the behaviour of molecules in the framework of quantum mechanics.

Thus, the problem is not to win an argument in favour of a particular methodological point of view; but to profit from criticisms in order to eliminate defects and inconsistencies from that point of view, and to show its potential applications.

We trust that the present study will be of use to those who are interested in assessing the reliability and the scope of approximate methods in terms of applications. We have tried to illustrate the potentialities and the limitations of those methods by concrete examples and to show that, if the underlying models are well understood, even simple methods like the Hückel one can still render invaluable services. We have also tried to duplicate as little as possible the existing excellent reviews of formalisms and applications. The present work should be considered complementary to the excellent study by Jug (1969), and largely complementary to several chapters that appeared in "Topics in Current Chemistry", especially those by Klopman and O'Leary (1970) and by Kutzelnigg et al. (1971), and to the review by Flanigan et al. (1977).

Another aspect of the present study that makes it somewhat special is that an attempt has been made to refer also to problems which are of interest to solid-state physicists, in particular those working on cluster calculations.

A larger part of the general theoretical analysis is directly derived from the lectures delivered by one of us (G.D.R.) to the "Herbstschule in Quantum Chemistry" organized in November 1976 by the Chemical Society of the DDR, at Kühlungsborn, East Germany: the stimulation that invitation provided is hereby acknowledged. Another source of encouragement was the Round Table meeting on the Ends and Means of Quantum Chemistry, organized by one of us (G.B.) in Obersteigen (France) in October 1976 under the sponsorship of the CNRS. There, all the participants agreed that work in the direction of the present study was badly needed, especially by those applying quantum chemical calculations.

Organic chemists, molecular biologists, and other users of theoretical methods should beware of uncritical escalation toward 'better' and 'better' methods, whether *ab initio* or semiempirical. It is much more important to know what information can be gained with and how reliable it is in terms of consistency with quantum mechanics. For these reasons, one should select the method which gives *the greatest insight* into one's physical or chemical problem, keeping in mind that a qualitative rule covering many data is far better than a very good quantitative result which gives no hint as to general trends.

This is why we have not given any special preference to the newest methods, and indeed have also devoted a large space to comparatively old results, especially when we could comment critically on them from our own experience.

Table of Contents

Chapter 1. Models and Concepts in Molecular Theory

1.1. Rôle of the Quantum Theory of Molecules

The work we shall discuss was designed to describe the rules governing molecular properties and resulting from applying the principles of quantum mechanics to their constituents atoms.

The most immediate examples are valency and aromaticity. In the former case, the well known rule that atoms form characteristic numbers of bonds must be related to the general principles of theoretical physics; in the latter case, the fact that a group of very reactive double bonds forms a stable, relatively unreactive entity must be explained in terms of a general rule derived from quantum mechanics.

The stability of the hydrogen molecule-ion and the non-existence of the helium molecule were explained by the general quantum law that two identical interacting systems have states which come in pairs, one lying above and the other below the pairs of degenerate states associated with the non-interacting systems; this is a result of 'coupling' of states coming from interaction. In the simple molecular orbital scheme, the explanation of the stability of the hydrogen molecule-ion is that an electron in the field of two protons has the energy of the lower one of the two states which derive from the coupling of the *1s* states of the electron on either nucleus; the explanation of the nonexistence of the helium molecule is that four electrons would 'occupy' both the bonding and the antibonding orbital associated with the same pair of atomic orbitals.

Let us consider more generally the theoretical concepts used in the traditional interpretation of the electronic and geometrical structure of molecules. The more or less intuitive notions of systematic chemistry in use, such as electronegativity, hybridization, etc. were developed during the thirties: the names of Hückel, Mulliken, Pauling, Van Vleck and other important scientists are associated with those years. Until very recently, there has been a tendency to believe that those concepts have been developed just because at the time it was almost hopeless to try to apply to larger molecular edifices numerical methods like those devised for the hydrogen molecule by Heitler and London and their followers. In recent years, however, it has become increasingly evident that the qualitative concepts of theoretical chemistry have still an important role to play in *rationalization* of both experimental data and quantum theoretical calculations. In spite of all the advances that have been made, they remain indispensable for the comprehension of molecular phenomena.

Certain theoretical schemes, which have also given rise to semiempirical methods, have played precisely the rôle of simplified physical models in the elaboration of those concepts. This consideration provides the soundest (and perhaps the only) justification for use of those semi-empirical methods at a time when ab-initio methods are considered more reliable black-box techniques for molecular calculations.

1.2 Born-Oppenheimer States

A molecule is a system of nuclei and electrons. Therefore, in the quantum mechanical description of the world that is the basis of present-day physics, its properties are determined by the motions of both nuclei and electrons: in other words, any interpretation of chemical properties must take into account both the electronic and the vibrorotational states of the molecule. An example is the chemical bond itself: if is impossible to explain the chemical bond without associating with it a force constant, i.e. without taking into account its properties with respect to vibrations.

Nevertheless, the Born-Oppenheimer approximation makes it possible to replace the overall molecular states by sets of electronic states where the nuclear positions enter as parameters - viz. the motion of the nuclei is considered to be infinitely slow. The principles of that approximation are well known. The total molecular (spin-independent non-relativistic) Hamiltonian $\hat{H}_{mol}$ is divided into three parts:

$$\hat{H}_{mol} = \hat{H}_{nn} + \hat{H}_{en} + \hat{V}_{en} \tag{1.1.1}$$

where

$\hat{H}_{nn}$ contains the nuclear kinetic energy $\hat{T}_{nn}$ and the nuclear repulsion $\hat{R}_{nn}$

$\hat{H}_{ee}$ contains the electronic kinetic energy $\hat{T}_{ee}$ and the electron repulsion $\hat{V}_{ee}$

$\hat{V}_{en}$ contains the interactions between electrons and nuclei.

Consider now states $|K\rangle$, $|\Phi\rangle$ constructed according to the equations

$$(\hat{H}_{ee} + \hat{V}_{en})|\Phi\rangle = \hat{E}_{nn}^{(\Phi)}|\Phi\rangle \tag{1.1.2}$$

$$(\hat{H}_{nn} + \hat{E}_{nn}^{(\Phi)})|K\rangle = E_{K\Phi}|K\rangle \tag{1.1.3}$$

where K refers to the M nuclei, Φ to the m electrons; $\hat{E}_{nn}^{(\Phi)}$ is a potential energy multiplier depending on the nuclear coordinates because the latter appear in $\hat{V}_{en}$; E is an energy.

The global states $|K, \Phi\rangle$ defined by Eqs. (1.1.2) and (1.1.3) constitute a basis in the Hilbert space associated with H_{mol}; but to consider one of them as a stationary state of a molecule is an approximation, - the Born-Oppenheimer (B-O.) approximation - because Eqs. (1.1.2) and (1.1.3) are not equivalent to the equation for the eigenstates of $\hat{H}_{mol}$. The quasi-stationary states $|K, \Phi\rangle$ are called B-O. states.

The essential point in the B-O. approximation as presented here is that so long as one is not interested in the nuclear states (vibrations and rotations) one can forget the nuclear part of the states under consideration and just consider electronic states satisfying Eq. (1.1.2) for given nuclear configurations. The multiplier $\hat{E}_{nn}^{(\Phi)}$ can be taken as the energy if the nuclear state corresponds to nuclei localized in given points of space (represented by a multidimensional vector $\vec{R}$). So far, it has been a current practice of much quantum chemical research to discuss only the electronic states $|\Phi(\vec{R})\rangle$ assuming that the nuclei are frozen in the equilibrium configuration $\vec{R}_0$. Recent interest in geometries and conformations has led to consideration of the states $|\Phi(\vec{R})\rangle$ associated with different nuclear configurations; indeed at least their energies are needed to study vibrational states in the B-O. approximation.

1.3 Computational and Interpretational Problems

It is well known that Eq. (1.1.2) is a complicated equation normally to be solved by numerical methods in a 3n-dimensional space - n being the number of electrons. Leaving aside the problem of determining and putting those results in a tractable way, in principle there are two possible ways of proceeding from Eq. (1.1.2). One can try to compute $|\Phi\rangle$ as accurately as possible, with the prospect of reconstructing the chemistry from it in the same way as was done on the chemical formula during the last century; or one can try to start directly coordinating the traditional theory of chemistry with quantum mechanics by taking shortcuts.

We shall not enter the discussion of the comparative merits of the two points of view. Suffice it to say that approximations are unavoidable even in the most accurate computations, at least for molecules of chemical interest: B.-O. approximation, truncation of the basis, neglect of correlation, neglect of relativistic effects. Moreover, at the interpretation stage, further approximations and simplifications must be made anyway.

One point of view amounts to demanding that quantum chemical computations should predict directly and quantitatively the properties of any specific molecule; it is valid if the goal is just to replace experiment by computation, at least in the case of very labile species, or of helping with such problems as the assignment of spectroscopic transitions; but it is not acceptable, at least in an extreme form, if the goal is interpretation and explanation in terms of simple quantum rules.

The kind of analysis involved in the interpretational approach is best explained by referring to a classical example from physics: real oscillators. There are two ways of dealing with problems involving those systems: Fourier analysis of the actual motion, and models - i.e. idealizations progressively approaching the real system by including finer and finer details. The two ways are perhaps complementary, but the one that has led theoretical mechanics to its greatest successes is the latter: The law of isochronism of a pendulum is based on an idealization.

Another impressive case where simplifications imposed by experimental apparatus have facilitated great discoveries was the spectrum of hydrogen: one wonders if, had a modern spectrograph been used, the simple regularities summarized in the Ritz principle would have been discovered so quickly.

The interpretational task of quantum chemistry can be fulfilled along paths parallel to the ones just mentioned for oscillators: idealizations and analysis of complete state vectors in a hierarchy of terms. The latter kind of procedure is widely applied in a posteriori analysis of accurate *ab initio* calculations, as in certain studies on localizability and transferability; the former is the basis of the scientifically most significant semi-empirical procedures.

1.4. Introduction of Simplified Models

To proceed gradually, let us consider first of all the well known example of the hydrogen molecule and its binding energy.

The steps can be listed as follows:

1) four (orthogonal) spin orbitals

$$|A\uparrow\rangle \quad |A\downarrow\rangle \quad |B\uparrow\rangle \quad |B\downarrow\rangle \tag{1.4.1}$$

representing one-electron states on atoms A and B with spins $\uparrow$ and $\downarrow$ are chosen as a basis;

2) the basis spin orbitals are combined to take account of the existence of two nuclei and of two electrons in one of two alternative ways:

a - (I) The atomic spin orbitals with the same spin associated with different atoms (which are coupled because an electron on B is subjected to the field of A and vice versa) are allowed to mix so as to give new one-electron states, the molecular spin orbitals:

$$
\begin{aligned}
|+\uparrow\rangle &= \frac{1}{\sqrt{2}}\left(|A\uparrow\rangle + |B\uparrow\rangle\right), \\
|-\uparrow\rangle &= \frac{1}{\sqrt{2}}\left(|A\uparrow\rangle - |B\uparrow\rangle\right), \\
|+\downarrow\rangle &= \frac{1}{\sqrt{2}}\left(|A\downarrow\rangle + |B\downarrow\rangle\right), \\
|-\downarrow\rangle &= \frac{1}{\sqrt{2}}\left(|A\downarrow\rangle - |B\downarrow\rangle\right).
\end{aligned}
\qquad (1.4.2)
$$

(II) Two-electron antisymmetrized states are built with the molecular orbitals (1.4.2):

$$|+\uparrow, +\downarrow\rangle, \ |+\uparrow, -\downarrow\rangle, \ |-\uparrow, +\downarrow\rangle, \ |-\uparrow, -\downarrow\rangle \qquad (1.4.3)$$

(These states are represented by Slater determinants when the wave function representation of states is adopted); and the lowest-energy state $|+\uparrow, +\downarrow\rangle$ is adopted as an approximate representation of the ground state of the hydrogen molecule. It is easy to prove that:

$$|+\uparrow, +\downarrow\rangle = \frac{1}{2}\left[|A\uparrow, A\downarrow\rangle + |A\uparrow, B\downarrow\rangle + |B\uparrow, A\downarrow\rangle + |B\uparrow, B\downarrow\rangle\right]. \qquad (1.4.4)$$

b - The two-electron states where one electron is on atom A and the other on atom B are combined linearly so as to obtain new two-electron states

$$|12, \uparrow\downarrow\rangle = \frac{1}{\sqrt{2}}\left(|A\uparrow, B\downarrow\rangle \pm |B\uparrow, A\downarrow\rangle\right). \qquad (1.4.5)$$

The energy lowering is then associated with the coupling of the degenerate two-electron states where the two electrons are assigned one to an atom, one to the other, with opposite spins.

The two approaches, which correspond to the molecular orbital (MO) method and to the Heitler-London (HL) method, respectively, are evidently not equivalent because of the presence in (1.4.4) of 'ionic' terms where the electron pair is localized on atom A or B. They are limiting cases of the more general expression (for the singlet state)

$$|\uparrow,\downarrow\rangle = c_1|A\uparrow, A\downarrow\rangle + c_2|B\uparrow, B\downarrow\rangle + c_3(|A\uparrow, B\downarrow\rangle + |A\downarrow, B\uparrow\rangle) \tag{1.4.6}$$

which corresponds to complete configuration interaction (CI) for our simple case. Now, in the molecular orbital scheme, binding is attributed to coupling of one-electron states, and the hydrogen molecule is treated as a sort of *double* one-electron molecule like H_2^+. In the HL bond scheme it is the indistinguishability of states where electrons are interchanged - and hence exchanged - that determines binding. The CI scheme is a combination of both.

The simple example of H_2 shows that approximate treatments may correspond to different ideas about what effect is the most important one in a given case. Of course, Eq. (1.4.4) somehow incorporates Eq. (1.4.5), so that it may be said that the MO scheme incorporates the HL scheme, but c_1 and c_2 of Eq. (1.4.4) - which are equal to each other in the case of H_2 - must be made equal to c_3. The MO method is thus no more general than the HL scheme.

In the MO scheme the binding is determined by the 'coupling' γ between atomic orbitals of different atoms, in the HL scheme it is determined by the coupling J between equivalent two-electron states:

$$J = \langle A\uparrow, B\downarrow|\hat{H}|A\downarrow, B\uparrow\rangle . \tag{1.4.7}$$

The quantities γ and J are the essential parameters of the two models: in the MO scheme, it is assumed that a proper choice of γ will lead to a consistent first-order description of the chemical bond in homonuclear diatomics; the equivalent assumption is made for J in the HL approach. Both γ and J can be regarded as *semiempirical parameters* and determined from experimental data: this does not affect the significance of the approach, because it is one of many ways of determining those values numerically. The validity of the corresponding approach must be judged on the overall picture it gives of the various physico-chemical properties of the bond.

1.5 Simplified Models and Model Hamiltonians

In the H_2 case just discussed we can take Eq. (1.4.6) as the complete scheme and Eq. (1.4.4) and (1.4.5) as simplified models, and the latter evidently correspond to limiting cases. They provide a partial answer to the question: To what extent can the binding in diatomics be treated as the result of just one effect - either coupling of one-electron states or exchange associated with mixing of two-electron wave functions? Both effects are general quantum mechanical ones: indeed they are the same general effect (splitting of degenerate states by perturbation) studied with different coupling schemes. Thus, if the answer to the above question is that one scheme, say the MO one, is particularly satisfactory, we can formulate a rule to be used as a guideline: From the point of view of quantum mechanics, the major contribution to covalent binding is provided by the coupling of atomic orbitals.

Assuming that the MO scheme is the better scheme, one may note that an important feature of the model is the number and the kind of atomic orbitals that are chosen. For instance, as we all know, significant improvements may be introduced by enlarging the basis so as to allow, among other things, for polarization. This means that in order to assess the validity and significance of a simplified scheme one must be very careful about its definition - and this is not an easy task: in fact, most of the misuse of semi-empirical methods derives from the introduction of improvements increasing their flexibility at the cost of serious inconsistencies.

A general definition of a simplified scheme where hidden faults are unlikely must correspond to a physical *model*. So, one considers an idealized physical system; in practice, *it must be possible to define a unique Hamiltonian operator* which satisfies all the rules of quantum mechanics and *represents the idealized system* in question. We shall base our whole study on this point.

The first example of a model we have in quantum chemistry results from the B.O. approximation; the molecule is replaced by a system of electrons in an external field, which is a physical system inasmuch as a perfectly acceptable Hamiltonian operator is associated with it.

A more complicated example of the rôle of simplified models, revealing the lights and shadows of a construction that was mostly developed on intuitive grounds, is the question of electronegativity. We shall treat it in detail because it is both instructive in itself and fundamental in connection with parameters.

1.6 Atoms in Molecules and Electronegativities

The idea of relating characteristic properties to the atomic constituents of a molecule is natural, but it is difficult to translate it into a well defined concept. One way to do that consists in associating with individual 'atoms in molecules' more or less strong electron-attraction properties: If two different atoms A and B are linked together by an electron pair, it is easy to admit that the electron cloud to which that electron pair may be assimilated is polarized at one end: A is said to be more electronegative than B if the center of (electrical) mass of the electron pair is shifted toward A; and vice versa.

The valence-state theory developed by Van Vleck and Mulliken in the years 1933-34 culminated in Moffitt's (1951) 'atoms-in-molecule' theory. But a repartition of the electrons among the various atoms remains necessary if atomic electrical charges are to be arranged so as to produce an electronegativity scale. Unfortunately, in spite of a few attempts (e.g. Parr et al. 1978), it has not been possible to obtain an 'absolute' scale, for the current definitions of electronegativity are based on quantum chemistry, which treats a molecule as a system of strongly interacting electrons; the electrons to be assigned to one of the 'atoms' of the molecule will not interact with one another and with 'their' nucleus much more than with the other electrons and the other nuclei of the same molecule. In other words, it has not been possible so far to isolate a part of the general molecular Hamiltonian that would represent the binding properties. Something of that kind has been done within the framework of the standard quantum chemical model approaches mentioned in Sec. 1.4. Of course, those approaches have produced different electronegativity scales; fortunately, they are compatible with each other because they are all based on the same building blocks, the atomic orbitals (AO's).

1.6.1. Electronegativity in the VB method. - The general CI expression (1.4.6) is the starting point of the valence-bond (VB) method for polyatomic molecules. As has been mentioned, its three terms correspond to 'ionic' and 'covalent structures'.

The orbitals $|A\rangle$, $|B\rangle$ with which the spin orbitals $|A\uparrow\rangle$, $|A\downarrow\rangle$, $|B\uparrow\rangle$, etc. are constructed may be either *pure* or *hybrid* AO's: The former case is illustrated by the H_2 molecule, where the best AO's for the simplest description are 1s orbitals; the latter case is illustrated by diamond, where sp^3 hybrids of carbon provide a very good localized description. As is well known, the coefficients c_1, c_2, c_3 of Eq. (1.4.6) are obtained from a secular equation involving the elements of the Hamiltonian matrix and of the overlap matrix associated with the three basis states $|A\uparrow, A\downarrow\rangle$, $|B\uparrow, B\downarrow\rangle$, $|12, \uparrow\downarrow\rangle_+$ (of Eq. (1.4.5)), and describe three states of the AB molecule (the ground state and two excited): for each state their square moduli measure the weights of the three basis states in $|\uparrow\downarrow\rangle$.

Pauling (1932) proposed that the two atoms A and B should be said to have the same electronegativity if $|A\uparrow A\downarrow\rangle$ and $|B\uparrow B\downarrow\rangle$ have the same coefficients - i.e. if the weights of the structures A^-B^+ and A^+B^- are the same in the ground state of the bond AB.

Next, the energies E_{AA} and E_{BB} for the two homonuclear molecules AA and BB were considered; and it was assumed that the purely covalent part of the energy E_{AB} of AB was given by

$$E_{A\text{-}B} = \text{mean of } (E_{AA}, E_{BB}) \ . \qquad (1.6.1)$$

Finally, the difference between the electronegativities X_A and X_B of the atoms forming the molecule AB was defined as

$$|X_A - X_B| = \sqrt{\frac{E_{A-B} - E_{AB}}{m}} \qquad (1.6.2)$$

where it is expected that E_{A-B}, being part of the total energy E_{AB}, is higher than E_{AB} (m is a constant).

Equation (1.6.2) produces an electronegativity scale provided the scale factor m and one electronegativity value are given: Pauling took $X_H = 2.1$ for hydrogen, and m = 23 kcal/mole.

It is well known that the evaluation of E_{A-B} created all sorts of difficulties; they arose from the fact that Eq. (1.6.1) is a purely intuitive one, and in fact the arithmetic mean was found to produce in some cases negative differences between E_{A-B} and E_{AB}. Geometric means (Pauling 1940) and harmonic means (Allen 1957) were proposed, but difficulties remained as is illustrated by the example of KH reported in Table 1. Those difficulties and the introduction of more sophisticated scales based on spectroscopic data eventually made the Pauling electronegativity scale somewhat obsolete.

In the present context, two points are instructive. First of all, the original idea of Pauling was based on a model where a diatomic molecule was interpreted as a three-state two-electron system, the three states being built on a four-dimensional basis.

Table 1.1. Bond Energy and Electronegativity Difference for the Molecule KH

Expression of E_{A-B}	$\Delta = E_{A-B} - E_{AB}$
$E_{A-B} = 1/2(E_{AA} + E_{BB})$	-15.1 kcal/mole
$E_{A-B} = -\sqrt{E_{AA} E_{BB}}$	6.5 -
$E_{A-B} = \frac{E_{AA} E_{BB}}{(E_{AA} + E_{BB})/2}$	20.2 -
$m(X_A - X_B)^2$	38.9 -

$D_{HH} = 104.2$ $D_{KK} = 13.2$ $D_{KH} = 43.6$ (kcal/mole^{-1})

$X_H = 2.1$ $X_K = 0.8$ (from Pauling 1940)

The latter consists of the four spin orbitals (1.4.1) that can be obtained from the orbital $|A>$ of A and the orbital $|B>$ of B. The model thus constructed can describe at most three (singlet) molecular states, of which the one with the lowest energy is expected to approximate the ground state of the real molecule in a satisfactory way. There are several conditions for the qualification 'satisfactory' to really apply. One of them is that it should be useful in correlating the properties of molecules with their structures - in particular the atoms entering their chemical structure. In other words, the theoretical model must represent correctly the changes in properties when one kind of atom is replaced by another in a given molecule. For one thing, this requires that the atomic orbitals $|A>$ and $|B>$ do not depend on the particular molecule in which they enter; and this is certainly not true at least if hybridization is admitted; even if we were interested just in diatomic molecules we should have difficulties with the so-called 'secondary' hybridization (Moffitt, 1950). Thus, Eq.(1.6.2) must be expected to give results valid under the assumption that atoms in molecules are represented by standard average orbitals, and it would be theoretically unsound to force it to do anything else. This is the explanation of the failures and the contradictions encountered in attempts to replace theoretical quantities by experimental data. Whereas shifts in zero point are allowed, so that E_{AB} and E_{A-B} can be replaced by dissociation energies (with negative signs), Eq.(1.6.1) is an empirical relationship which cannot hold as such, because (i) the experimental energies of homonuclear diatomics contain ionic contributions, and (ii) the properties of the atomic orbital best qualified to represent a given atom in a molecule depend on its partner. A step toward a greater consistency of the model used by Pauling was introduced by Mulliken (1934). Let H_{ij} and S_{ij} denote the elements of the Hamiltonian matrix and of the overlap matrix over the three states appearing in Eq.(1.4.6), and E be the lowest eigenvalue of the Hamiltonian matrix. The elementary theory of homogeneous linear systems then implies that the coefficients c_1, c_2, c_3 of Eq.(1.4.6) are proportional to the minors formed by the first two rows and by the pairs of columns (2,3), (3,1), (1,2), respectively, of the matrix W whose general element is

$$W_{ij} = H_{ij} - ES_{ij} = W_{ji} \quad . \tag{1.6.3}$$

Now, as has been mentioned, if two atoms A, B have the same electronegativity, the ground (lowest-energy) state of the type (1.4.6) must have $c_1 = c_2$; in terms of the matrix W this means

$$\det\begin{vmatrix} W_{12} & W_{13} \\ W_{22} & W_{23} \end{vmatrix} = \det\begin{vmatrix} W_{13} & W_{11} \\ W_{23} & W_{21} \end{vmatrix} \quad . \tag{1.6.4}$$

If, by analogy to what happens in homonuclear molecules it is also assumed that $W_{13} = W_{23}$ (the coupling terms of the ionic structures with the covalent one are equal), Eq.(1.6.4) reduces to

$$W_{22} = W_{11}, \quad \text{i.e.} \quad H_{22} = H_{11} \tag{1.6.5}$$

(S_{jj} being unity). The equality of the energies H_{11} and H_{22} associated with the two ionic terms implies

$$I_B - A_A = I_A - A_B \quad \text{or} \quad I_B + A_B = I_A + A_A , \tag{1.6.6}$$

where I and A stand for ionization potentials and electron affinities: the equality in weight of the ionic terms amounts to equality of the means of ionization potentials and of electron affinities. Therefore, considering that (1.6.6) refers to two electrons, the electronegativity of A can be defined as

$$\chi_A = \frac{1}{2} (I_A + A_A) . \tag{1.6.7}$$

An absolute electronegativity scale has thus been obtained and the dependence on experimental dissociation energies has been removed. This result, plus the fact that Mulliken's scale (with a few exceptions) parallels that of Pauling, are important points in favour of the model and of the electronegativity concept. Moreover, the problem of determining the 'covalent bond energy' has disappeared.

Nevertheless, two major theoretical objections remain, one internal, the other external to the model. First, the assumption has been made that even in a heteronuclear diatomic molecule formed by atoms having the same electronegativity the coupling between *either* ionic basis state and the covalent basis state is the same; second, the atoms are still represented by a single standard orbital. The former assumption demands a study outside the scope of the present work; the latter is a limitation that can be removed by introducing an extended version of the model.

1.6.2. The valence state concept. - The definition (1.6.7) of electronegativities does not depend on the condition that the atomic orbital of a given atom is always the same - provided we know how to define a ionization potential and an electron affinity for a given orbital of an atom in its *valence state*.

The concept of valence state derives from the Slater atomic model, widely used in spectroscopy and itself an adaptation of the electron shell model which allows separation of the spectroscopic components of a degenerate atomic configuration (e.g. the states ^1S, ^3P, ^1D of s^2p^4 oxygen).

The quantities essential to that model are the Slater atomic parameters: W_i, $F_{n,\hat{i}}$ (n = 0, 2, 4, ...), $G_{m,ij}$ (m = 1, 3 ...); they provide either the interaction energy I_i of the valence electron occupying the i-th orbital with the nucleus and the inner electrons or the spherical-harmonic expansion of two-electron integrals J_{ij} (Coulomb) and K_{ij} (excharge) associated with the orbital pair $|\chi_i\rangle$, $|\chi_j\rangle$. (The subscripts i, j correspond to the s, p, d, ... orbitals of the outer shell.) With the exception of the parameters W_i and F_0, evaluation of which involves ionization potentials, the quantities in question are obtained from experimental spectroscopic data for atom A in its ground-state configuration, but least-square adjustments are normally necessary.

The valence state is defined by a sort of limiting process: the interatomic distance is taken to infinity without changing the spin-coupling of the bonding electrons. In the case of a covalent two-electron bond between A and B, cleavage of the AB bond leaves one electron on A (and one on B), with a random orientation of its spin with respect to the other electrons of A (or of B). An atom containing n valence electrons is said to be in the V_n valence state; the corresponding total energy $E(V_n)$ is obtained by counting the interaction terms I_i and J_{ij} in the usual way, and by taking the average exchange contribution $\frac{1}{2}K_{ij}$ for the contribution to exchange of random-spin electrons:

$$E(V_n) = \sum n_i I_{\hat{i}} + \sum (n_i - 1)J_{ii} + \sum_i \sum_{j<i} n_i n_j (J_{ij} - \frac{1}{2} K_{ij}) \tag{1.6.8}$$

where n_i and n_j are the occupation numbers (2, 1, or 0) of the atomic orbitals $|\chi_i\rangle$ and $|\chi_j\rangle$. As the integrals I_i, J_{ij}, K_{ij} are computed from the Slater-Condon parameters as in the case of the free atom, the energy $E(V_n)$ can be expressed as a weighted sum of spectroscopic state energies, even though the valence state is not an observable atomic state.

As is well known, Mulliken electronegativities for valence states have been calculated for many atoms; for light atoms from lithium to chlorine (Pilcher and Skinner 1962, Hinze and Jaffé 1962), for the transition metals of the iron family (Hinze and Jaffé, 1963), for the elements of the other long periods (Di Sipio et al. 1971). An abstract is given in Table 2.

The electronegativity changes observed for low energy orbitals like the 2p orbital in the C, N, O series are mainly associated with the changes in the first ionization potentials, for the latter are generally much greater than electron affinities (e.g., I = 10 ev, A = 1 ev). (In molecules containing certain heavy atoms, it can happen that an atom having a weak ionization potential is linked to an atom of which the electron affinity has the same order of magnitude; the standard example is cesium fluoride $C_s^+F^-$: I_{Cs} = 3.9 ev, A_F = 3.5 ev).

Table 1.2. Mulliken and Pauling Electronegativities of Selected Atoms in Various Valence states. (Hinze and Jaffé 1962, 1963a, 1963b)

		I(eV)	A(eV)	χ_M	χ_P	o)
H (V_1, s)	s	13.60	0.75	7.17	2.23	(2.1)
C $(V_4, sppp)$	s	21.01	8.91	14.96	4.88	(2.5)
	p	11.27	0.34	5.80	1.76	
N (V_3, s^2ppp)	p	13.94	0.78	7.36	2.29	
N (V_3, sp^2pp)	s	26.92	13.99	20.46	6.75	(3.0)
	p	14.42	2.54	8.48	2.67	
O (V_2, s^2p^2pp)	p	17.28	2.01	9.65	3.07	(3.5)
O (V_2, sp^2p^2p)	s	36.07	18.44	27.25	9.05	
	p	18.53	3.40	10.96	3.52	
F $(V_1, s^2p^2p^2p)$	p	20.86	3.50	12.18	3.93	(4.0)
F $(V_1, sp^2p^2p^2)$	s	38.24	24.37	31.30	10.43	
Si $(V_4, sppp)$	s	17.31	6.94	12.12	3.91	(1.8)
	p	9.19	2.82	6.00	1.83	
P (V_3, s^2ppp)	p	10.73	1.39	6.06	1.85	
P (V_3, sp^2pp)	s	20.20	8.45	14.32	4.66	(2.1)
	p	12.49	1.98	7.23	2.25	
S (V_2, s^2p^2pp)	p	12.39	2.38	7.38	2.30	(2.5)
S (V_2, sp^2p^2p)	s	20.03	11.54	15.78	5.15	
	p	13.32	3.50	8.41	2.65	
Cl $(V_1, s^2p^2p^2p)$	p	15.03	3.73	9.38	2.98	(3.0)
Cl $(V_1, sp^2p^2p^2)$	s	24.02	14.45	19.23	6.33	
Ge $(V_4, sppp)$	s	18.57	6.86	12.71	4.11	(1.8)
	p	9.43	4.26	6.84	2.11	
As (V_3, s^2ppp)	p	9.36	1.33	5.34	1.60	
As (V_3, sp^2pp)	s	16.22	7.92	12.07	3.89	(2.0)
	p	12.16	3.38	7.77	2.43	
Se (V_2, s^2p^2pp)	p	11.68	2.52	7.10	2.20	(2.4)
Se (V_2, sp^2p^2p)	s	20.49	10.36	15.42	5.30	
	p	14.44	2.04	8.24	2.59	
Br $(V_1, s^2p^2p^2p)$	p	13.10	3.70	8.40	2.64	(2.8)
Br $(V_1, sp^2p^2p^2)$	s	22.70	14.50	18.28	6.00	
Sb $(V_4, sppp)$	s	16.16	7.72	11.94	3.85	(1.8)
	p	8.32	5.33	6.82	2.11	
Sb (V_3, s^2ppp)	p	8.75	1.18	4.96	1.48	
Sb (V_3, sp^2pp)	s	18.80	7.51	13.15	4.26	(1.9)
	p	11.68	3.62	7.65	2.39	
Te (V_2, s^2p^2pp)	p,	11.04	2.58	6.81	2.10	(2.1)
Te (V_2, sp^2p^2p)	s	20.78	9.09	14.93	4.87	
	p	14.80	2.93	8.86	2.80	
I $(V_1, sp^2p^2p^2)$	s	18.00	13.38	15.69	5.12	
I $(V_1, s^2p^2p^2p)$	p	12.67	3.52	8.09	2.54	(2.5)

o) Values in parentheses are Pauling electronegativities.

Table 1.3a. Same as Table 1.2 for hybrid orbitals *

Type mult. bond	χ_M s or σ	 p or π	 single bond	χ_M s or σ	 p or π
			Carbon V_4		
s p p p	14.96	5.80			
di di	10.38	5.64			
tr tr tr π	8.79	5.59			
			te te te te	7.75	
			Nitrogen V_3		
s^2p p p		7.36	s p^2p p	20.46	8.48
di^2di π π	15.64	7.92	di di $\pi^2\pi$	14.47	8.12
tr^2tr tr π	12.87	7.95	tr tr tr π^2	12.31	
			te^2te te te	11.54	
			Oxygen V_2		
s^2p^2p p $di^2di^2\pi$ π		9.64			
s p^2p^2p	27.25	10.96			
di^2di π^2 π	20.20	10.30	di di $\pi^2\pi^2$	19.11	
tr^2tr^2tr π	17.07	10.08	tr^2tr tr π^2	16.73	
			te^2te^2te te	15.25	
			Fluorine V_1		
			s $p^2p^2p^2$	31.30	
			$s^2p^2p^2$p		12.18
			Silicon V_4		
s p p p	12.12	6.00			
di di π π	9.06	5.69			
tr tr tr π	7.90	5.58			
			te te te te	7.29	
			Phosphorus V_3		
s^2p p p		6.06	s p^2p p	14.32	7.23
di^2di π π	11.23	6.64	di di π^2 π	10.77	6.95
tr^2tr tr π	9.66	6.72	tr tr tr π^2	9.47	
			te^2te te te	8.90	

Table 1.3a (continued)

Type	χ_M			χ_M	
mult. bond	s or σ	p or π	single bond	s or σ	p or π
		Sulphur V_2			
$s^2p^2p\,p$ / $di^2di^2\pi\,\pi$		7.38			
$s\,p^2p^2p$	15.78	8.41			
$di^2di\,\pi^2\pi$	12.37	7.90	$di\,di\,\pi^2\pi^2$	12.11	
$tr^2tr^2tr\,\pi$	10.88	7.73	$tr^2tr^2tr\,\pi^2$	10.88	
			$te^2te^2te\,te$	10.13	
		Chlorine V_1			
			$s\,p^2p^2p^2$	19.23	
			$s^2p^2p^2p$		9.38
		Germanium V_4			
$s\,p\,p\,p$	12.71	6.84			
$di\,di\,\pi\,\pi$	9.78	6.51			
$tr\,tr\,tr\,\pi$	8.66	6.41			
			$te\,te\,te\,te$	8.07	
		Arsenium V_3			
$s^2p\,p\,p$		5.34	$s\,p^2p\,p$	12.07	7.77
$di^2di\,\pi\,\pi$	9.01	6.55	$di\,di\,\pi^2\,\pi$	9.92	7.72
$tr^2tr\,tr\,\pi$	8.53	6.94	$tr\,tr\,tr\,\pi^2$	9.18	
			$te^2te\,te\,te$	8.30	
		Selenium V_2			
$s^2p^2p\,p$ / $di^2di^2\pi\,\pi$		7.10			
$s^2p^2p\,p$	15.42	8.24			
$di^2di\,\pi^2\,\pi$	11.86	7.67	$di\,di\,\pi^2\pi^2$	11.83	
$tr^2tr^2tr\,\pi$	10.41	7.48	$tr^2tr\,tr\,\pi^2$	10.52	
			$te^2te^2te\,te$	9.76	
		Bromine V_1			
			$s\,p^2p^2p^2$	18.28	
			$s^2p^2p^2p$		8.40

Table 1.3a (continued)

Type mult. bond	χ_M s or σ	χ_M p or π	single bond	χ_M s or σ	χ_M p or π
		Tin V_4			
s p p p	11.94	6.82			
di di π π	9.39	6.55			
tr tr tr π	8.40	6.45			
			te te te te	7.89	
		Antimonium V_3			
s^2p p p		4.96	s p^2p p	13.15	7.65
di^2di π π	9.81	6.31	di di π^2 π	10.40	7.38
tr^2tr tr π	8.93	6.64	tr tr tr π^2	9.37	
			te^2te te te	8.47	
		Tellurium V_2			
s^2p^2p p } di^2di^2 $\pi\pi$		6.81			
s p^2p^2p	14.93	8.86			
di^2di $\pi^2\pi$	11.48	7.83	di di $\pi^2\pi^2$	11.90	
tr^2tr^2tr π	10.05	7.49	tr^2tr tr π^2	10.47	
			te^2te^2te te	9.65	
		Iodine V_1			
			s $p^2p^2p^2$	15.69	
			$s^2p^2p^2$p	8.09	

* te = tetrahedral hybrid; tr = trigonal hybrid; di = digonal hybrid; π = π-type p orbital.

1.6.3. Hybridization and Electronegativities. - As can be seen in Table 3, the numbers obtained from valence state ionization potentials and electron affinities coincide with the electronegativities originally derived by Pauling in the cases where, according to the elementary orbital theory, the latter participate in bonds by almost pure orbitals (H and F); otherwise, they define a range within which Pauling's values lie: the latter then represent the electronegativities of orbitals intermediate between s and p (C, N, O).

The above consideration shows that the hybridization concept comes quite naturally into its own as the definition of electronegativity is made more and more accurate.

Table 1.3b. Same as table 1.2 for heavy atoms

Hybrid orbitals	I(eV)	A(eV)	χ_M	χ_A
		Chromium		
Tetrahedral sp^3	6.59	35.91	21.25	18.88
plane tetragonal dsp^2	8.63	20.69	14.66	4.77
Octahedral d^2sp^3	9.08	35.53	22.30	7.37
		Manganese		
Tetrahedral sp^3	12.12	48.23	30.17	10.05
plane tetragonal dsp^2	10.91	23.72	17.31	5.67
Octahedral d^2sp^3	16.34	44.81	30.57	10.18
		Iron		
Tetrahedral sp^3	31.82	21.19	26.55	8.82
plane tetragonal dsp^2	21.19	12.58	16.88	5.53
Octahedral d^2sp^3	39.08	19.78	29.43	9.80

In fact, it is possible to define hybrid-orbital electronegativities by combining linearly the ionization potentials and the electron affinities associated with 'pure' orbitals s, p, ... in the appropriate combinations of valence states (i.e. taking into account the s or p character of the orbital under consideration), when the pure orbitals are occupied in part by pairs of electrons. In general, the electronegativity of a σ orbital is very close to a linearly increasing function of the s-character ($(n+1)^{-1/2}$ in an sp^n hybrid), an almost trivial result which played a crucial role in the development of the well-known Walsh rules (1946-53).

Of course, a rigorous analysis of the hybridization concept and an unambiguous definition of the hybridization associated with a given bond is necessary at this stage. Indeed, the role of AO's - in particular of a truncated basis of (presumably) best AO's - in the theoretical model within which electronegativity and hybridization are introduced must be further examined.

Clearly the validity of those theoretical concepts does not result from the more or less empirical source of the quantitative estimates, but from the absence of contradictions and from consistency of definitions within the underlying model.

The valence-state problem and the role of hybridization show that the quest for a quantitative and consistent definition of electronegativity is, to say the least, a very fruitful line of work. The various features of elementary quantum chemistry come into play and one after the other demand a more rigorous definition.

In fact, many inconsistencies stem from the use of experimental data involving the entire reality, and not a simplified model; the case of Pauling's estimates of the covalent-bond energy illustrates the kind of difficulties that can arise in this connection. One would be on safe grounds only if it were sure that the 'details' of reality not included in the model are negligible as concerns the experimental values to be used. Otherwise, it may be more reasonable to adopt a completely theoretical scheme, as has in fact been done by a number of authors (Mulliken 1935a, b; Sanderson 1974 etc.).

We shall come back to electronegativities in connection with the purely theoretical approach after we have introduced the independent-electron model.

1.7 A Conclusion: Models and the Plague of Non-observables

In the present chapter we have tried to present the notion of a theoretical model and to connect it with the introduction of special concepts. We have carried out a detailed discussion on the example of electronegativity, which is the source of many parameterizations for molecules and solids. Concepts like net atomic charges, bond orders, etc. are also of great importance and can be defined and discussed along the same lines (cf. Del Re and Lami, 1976; Barone et al., 1979). We shall confine ourselves to the following general consideration.

Any approach to chemistry that is not merely descriptive must involve some classification and some rules. In turn, the latter demand suitable concepts which may be to some extent artificial. The most famous example is aromaticity, which has given rise to a number of daughter concepts, like pseudo-aromaticity, anti-aromaticity, etc. (Bergman and Pullman, 1973). The usefulness of such concepts led to abuse; abuse caused a reaction against "the plague of non-observables".

The argument behind that expression runs as follows. An essential feature of quantum mechanics is that only equations involving observable quantities should be written, even if that requires accepting a certain degree of ignorance (as is expressed in the uncertainty principle). Therefore, quantities such as net atomic charges, chemical bonds, and the like, which are not measurable, have no place in physics.

The argument, taken in its generality, is not acceptable. Every measurement is an indirect one, except measurements of length and mass (and perhaps time): a theory is needed to compute the quantity that has been measured from observed lengths, masses, and times. Therefore, any quantity may be classified as observable as long as a *unique* recipe for computing it from experimental observations is given. Such could be the case, for instance, with regard to bond properties: as long as the bond networks of molecules are uniquely defined, and if bond enthalpies are defined as standard quantities reproducing in the best possible way the enthalpies of formation of a suitable *uniquely* specified set of molecules, we have defined *observable* bond enthalpies. Of course it remains to be seen whether or not the resulting values are useful for describing other molecules or for formulating general rules.

In short, the real dangers in definitions of auxiliary concepts are twofold: the definitions may be ambiguous, and incorrect interpretations or applications may be suggested: one ought to speak of 'the plague of ill-defined and wrongly applied concepts', which is a very real plague in physics as well as in chemistry.

Chapter 2. Mathematical Foundations

2.1 Mathematical Construction of Many-Electron Models from an Orbital Basis

The traditional scheme of quantum chemistry involves a basis of one-electron states (orbitals) as the starting point of any treatment. Different points of departure are possible, but experience gained so far in that connection is very limited; therefore, it is reasonable that attention should be confined to orbitals as the building blocks of many-electron states on which models are defined (through the corresponding Hamiltonian operators).

Intuitively, to assume that a set of orbitals is given means that we have in mind a system similar to our molecule, but where all the electrons except one are either removed or melted into the nuclei, so as to appear just as contributions to the field sources. This is precisely the case when we define atomic one-electron states (the AO's); we consider fictitious limiting systems where all the electrons and nuclei of the given molecule have been taken to infinity except one nucleus and its electrons, and we treat the remaining atom as a one-electron system. If we assume that the atom is left in the ground state, then either SCF orbitals of the free atom or Slater orbitals can be the atomic one-electron states under consideration.

Many electron states are not necessarily built from AO's; indeed, current treatments involve first the computation of suitable molecular orbitals, and then the so-called CI (configuration interaction) step. Therefore, the only restriction which will be placed on the orbital basis in the following considerations will be that it forms an orthonormal set. That restriction greatly simplifies the general work, and can be eliminated in a straightforward way (Grimley 1967).

We now proceed to describe how a many-electron basis and a many-electron Hamiltonian are associated with that basis. The most direct way is through creation and annihilation operators, which have recently acquired full recognition in quantum chemistry (see e.g. Avery 1977).

2.1.1. Second quantization formalism. - We start from a set of one-electron states (*spin orbitals*) obtained from a given basis of p orthonormal orbitals. They will be denoted by a (Greek or Latin) letter specifying the orbital followed by s to denote the general spin state; s can be specified by $\uparrow$ (for spin α) or $\downarrow$ (for spin β). We can then write n-electron states $|M\rangle$, $|N\rangle$, in one of two ways:

$$1):\quad |M\rangle = |\varphi_1 s_1, \varphi_2 s_2, \ldots, \varphi_n s_n\rangle \qquad (2.1.1)$$

where the *occupied* spin orbitals are specified;

$$2):\quad |M\rangle = |n_{1\uparrow}^{(M)}\, n_{1\downarrow}^{(M)}\, n_{2\uparrow}^{(M)}\, n_{2\downarrow}^{(M)} \ldots n_{p\downarrow}^{(M)}\rangle \qquad (2.1.2)$$

where the 2p *available* spin orbitals are supposed to be known, and only the proper occupation number $n_{js}^{(M)}$ (0 to 1) of each $|js\rangle$ in $|M\rangle$ is given.

All the many-electron states (M.E.S.) are assumed to be antisymmetrized with respect to interchange of electron pairs. Because of the orthonormality of the spin orbitals they are orthogonal to one another. They can be thought of as Slater determinants obtained by occupying in various ways the available spin orbitals. For example, the wavefunction representing the state $|101101\rangle$ of the π electrons of the system

$$\ddot{X} - Y = Z \qquad \text{e.g.} \qquad NH_2 - CH = CH_2$$

may be written in the form

$$\varphi(101101) = \frac{1}{4!}\begin{vmatrix} p\pi_x\uparrow(1) & p\pi_x\uparrow(2) & p\pi_x\uparrow(3) & p\pi_x\uparrow(4) \\ p\pi_y\downarrow(1) & p\pi_y\downarrow(2) & p\pi_y\downarrow(3) & p\pi_y\downarrow(4) \\ p\pi_y\uparrow(1) & p\pi_y\uparrow(2) & p\pi_y\uparrow(3) & p\pi_y\uparrow(4) \\ p\pi_z\downarrow(1) & p\pi_z\downarrow(2) & p\pi_z\downarrow(3) & p\pi_z\downarrow(4) \end{vmatrix} \qquad (2.1.3a)$$

and represents the "resonance structure" where the four π electrons are localized according to the scheme

$$\overset{\uparrow}{X}{}^{\oplus}\ \overset{\uparrow\downarrow}{Y}{}^{\ominus}\ \overset{\downarrow}{Z} \qquad \text{e.g.} \qquad H_2\overset{+}{N} - \overset{-}{C}H - CH_2\ . \qquad (2.1.3b)$$

(We note explicitly that the word 'configuration' will be used for our M.E.S. when ambiguities are not dangerous (as in the expression 'configuration interaction', CI); but otherwise, it is better to think of configurations as of combinations of our M.E.S. that are eigenfunctions of S_z.)

Next, following a standard procedure, creation and annihilation (C.A.) operators $\hat{a}_{js}^{+}$, $\hat{a}_{js}$ are defined by the equations:

$$\hat{a}_{js}^{+}|n_{1\uparrow} n_{2\downarrow} n_{2\uparrow} \ldots n_{js} \ldots\rangle = (-1)^{N_{js}} \sqrt{1-n_{js}}\, |n_{1\uparrow} n_{1\downarrow} \ldots n_{js}+1 \ldots\rangle \qquad (2.1.4a)$$

$$\hat{a}_{js}|n_{1\uparrow}n_{1\downarrow}n_{2\uparrow}\ldots n_{js}\ldots> = (-1)^{N_{js}}\sqrt{n_{js}}\,|n_{1\uparrow}n_{1\downarrow}n_{2\uparrow}\ldots n_{js}-1\ldots> \qquad (2.1.4b)$$

where N_{js} is the number of electrons occupying spin orbitals listed before js in the given state. This means that $\hat{a}^+_{js}$ generates from any state with an empty js-th spin orbital a new state where that spin orbital is occupied; otherwise it destroys the state on which it operates. Similarly, $\hat{a}_{js}$ empties the js-th spin orbital if it is full. The signs are chosen so as to ensure consistency in successive applications of the operators in question.

The product

$$\hat{n}_{js} = \hat{a}^+_{js}\hat{a}_{js} \qquad (2.1.5)$$

is the *number operator* because it just multiplies the state on which it acts by the number of electrons occupying the js-th spin orbital; the operator

$$\hat{b}_{ks'\to js} \equiv \hat{a}^+_{js}\hat{a}_{ks'} \qquad (2.1.6)$$

may be called the ks'→js shift operator, because it acts on a state by shifting one electron from the ks'-th spin orbital to the js-th one, if possible, or destroys it.

In general, products of creation and annihilation operators in equal numbers leave the number of electrons unchanged, but shift electrons from one spin orbital to another. Examples referring to the system XYZ of Eq.(2.1.3b) are

$$\hat{a}^+_2\hat{a}_6|101101> = -\hat{a}^+_2|101100> = |111100>,$$

or

$$\hat{a}^+_2\hat{a}_6|\overset{\uparrow}{X}^{\oplus} - \overset{\downarrow\uparrow}{Y}^{\ominus} - \overset{\downarrow}{Z}> = -\hat{a}^+_2|\overset{\uparrow}{X}^{\oplus} - \overset{\downarrow\uparrow}{Y}^{\ominus} - Z^{\oplus}> = |\overset{\uparrow\downarrow}{X} - \overset{\downarrow\uparrow}{Y}^{\ominus} - Z^{\oplus}> \qquad (2.1.7a)$$

$$\hat{a}^+_2\hat{a}_4|101101> = \hat{a}^+_2|101001> = -|111001>$$

or

$$\hat{a}^+_2\hat{a}_4|\overset{\uparrow}{X}^{\oplus} - \overset{\downarrow\uparrow}{Y}^{\ominus} - \overset{\downarrow}{Z}> = \hat{a}^+_2|\overset{\uparrow}{X}^{\oplus} - \overset{\uparrow}{Y} = \overset{\downarrow}{Z}> = -|\overset{\uparrow\downarrow}{X} - \overset{\uparrow}{Y} = \overset{\downarrow}{Z}> \qquad (2.1.7b)$$

$$\hat{a}^+_5\hat{a}^+_2\hat{a}_4\hat{a}_6|101101> = -\hat{a}^+_5\hat{a}^+_2\hat{a}_4|101100> = -\hat{a}^+_5\hat{a}^+_2|101000> =$$

$$= +\hat{a}_5|111000> = -|111010>$$

or

$$\hat{a}^+_5\hat{a}^+_2\hat{a}_4\hat{a}_6|\overset{\uparrow}{X}^{\oplus} - \overset{\uparrow\downarrow}{Y}^{\ominus} - \overset{\downarrow}{Z}> = -|\overset{\uparrow\downarrow}{X} - \overset{\uparrow}{Y} - \overset{\uparrow}{Z}> . \qquad (2.1.7c)$$

Products of the C.A. operators for orthonormal bases obey the anticommutation rules:

$$\hat{a}^{+}_{js}\hat{a}^{+}_{ks'} + \hat{a}^{+}_{ks'}\hat{a}^{+}_{js} = 0 \tag{2.1.8a}$$

$$\hat{a}^{+}_{js}\hat{a}_{ks'} + \hat{a}_{ks'}\hat{a}_{js} = 0 \tag{2.1.8b}$$

$$\hat{a}^{+}_{js}\hat{a}_{ks'} + \hat{a}_{ks'}\hat{a}^{+}_{js} = \delta_{jk}\delta_{ss'} \tag{2.1.8c}$$

The C.A. operators can be used to generate all the n-electron states associated with a given set of spin orbitals from a single reference state $|M_0>$ (or even from the 'vacuum' by successive creation of electrons in the various spin orbitals). Now, the n-electron states (2.1.2) are assumed to form a complete basis in the state space of an n-electron system; every state of that system can be expressed as a linear combination of the states represented by Eq. (2.1.2). It is legitimate in principle and necessary in practice to truncate the M.E.S. basis exactly as is done with the generating one-electron basis. That truncation is part of the formulation of a specific many-electron model starting from the given one-electron basis; in other words, neglect of certain configurations is a feature of the special model, and need not be decided when constructing the many-electron basis.

In short, we shall work in the many-electron Hilbert space spanned by all the M.E.S. that can be built from the given (truncated) orbital basis; and we shall keep in mind that, as has been illustrated in Eqs. (2.1.7), the many-electron basis is generated entirely from one state of it by the appropriate products of equal numbers of creation and annihilation operators.

The latter remark is the starting point of the representation of linear operators as linear combinations of C.A. operator products.

We know that every linear operator takes a state into another state; in particular it takes any given basis state into a specific combination of basis states. We are interested here in the Hamiltonian operator $\hat{H}_{el}$. Now, it is well-known that (for orthonormal basis states)

$$\hat{H}_{el}|M> = \sum_{N} |N><N|\hat{H}_{el}|M> \tag{2.1.9}$$

where $<N|\hat{H}_{el}|M>$ is the matrix element of $\hat{H}_{el}$ over the M.E.S. $|N>$ and $|M>$, and is a scalar. Next, consider the operator $|N><M|$: it can be written as a product $\hat{\Pi}_{M\to N}$ of C.A. operators, because either it applies to $|M>$ to give $|N>$ or it gives zero

$$(|N><M|)|M'> = |N><M|M'> \equiv \hat{\Pi}_{M\to N}|M'> = |N>\delta_{NM'} \ . \tag{2.1.10}$$

Therefore, multiplying (2.1.9) by $<M|$ on the right, summing over M: and remembering that $\sum_M |M><M| = \hat{1}$, we can write:

$$\hat{H}_{el} = \sum_M \sum_N <N|\hat{H}_{el}|M> \hat{\Pi}_{M\to N}|M><M| \ . \qquad (2.1.11)$$

Finally, expansion of (2.1.11) into orbitals and individual C.A. operators shows that the action of the Hamiltonian operator amounts to generating from any given basis state of the type (2.1.1) the following states:

(i) the same state multiplied by a constant R_0;

(ii) all the states obtainable by emptying the ks'-th spin orbital and filling the js-th one, multiplied each by a coefficient $<js|\hat{H}_1|ks'>$;

(iii) all the states obtainable by emptying the ks'-th *and* the ms'''-th spin orbital, and filling the js-th *and* the ls''-th spin orbital, and multiplying each by a coefficient $\frac{1}{2}$ (jsks'|ls''ms'''), the factor $\frac{1}{2}$ being due to the fact that the sum is taken over all the values of the indices.

In general, there should be more terms, corresponding to interchanges of 3, 4, ... electrons. However, such terms bear vanishing coefficients when the potential energies appearing in the Hamiltonian correspond at most to two-particle interactions. Therefore, the Hamiltonian operator can be written in the form

$$\hat{H}_{el} = R_0 + \sum_{js,ks'} <js|\hat{H}_1|ks'>\hat{a}^+_{js}\hat{a}_{ks} + \frac{1}{2} \sum_{js,ks',cs'',ms'''} (jsks'|ls''ms''')\hat{a}^+_{js}\hat{a}^+_{ls''}\hat{a}_{ms'''}\hat{a}_{ks'} \ . \qquad (2.1.12)$$

Indeed, by introducing Slater determinants, it can be shown that

(i) R_0 is the constant part of $\hat{H}_{el}$

(ii) $<js|\hat{H}_1|ks'>$ is an element of the matrix representing one of the n identical one-particle terms of the Hamiltonian operator, $\hat{H}_1$, in the basis of the orthonormalized spin orbitals (...|js>...|ks'>...) from which the states (2.1.2) have been built;

(iii) $<jsks'|ls''m'''>$ is an element of the hypermatrix obtained from the two-particle operators

$$<ls''|\hat{V}_{12}|ms'''>$$

(which are operators because |ls''> and |ms'''> are states of one electron; say, electron 2) and then taking their matrix elements

$$<jsks'|ls''ms'''> = <js|<ls''|\hat{V}_{12}|ms'''>|ks'> \ . \qquad (2.1.13)$$

If $\hat{H}_{el}$ does not contain spin, s and s' must be equal when they refer to the same electron, because the scalar product of different spin states is zero. Therefore, unless Löwdin's DODS (different orbitals for different spins) ansatz is introduced, (2.1.12) can be written in the form

$$\hat{H}_{el} = R_0 + \sum_{js',ks''} H_{jk}\hat{a}^+_{js}\hat{a}_{ks} + \frac{1}{2} \sum_{\substack{js,ks' \\ ls'',ms'''}} (jk|lm)\hat{a}^+_{js}\hat{a}^+_{ls''}\hat{a}_{ms'''}\hat{a}_{ks'} \; . \quad (2.1.14)$$

where j, k, l, m *denote the orbitals associated with the corresponding spin orbitals:* j and l represent the same orbital, if the js-th and the ls'-th spin orbitals correspond to a doubly filled orbital and, of course, have opposite spins.

The essential parameters entering Eq. (2.1.14) are:

(i) the zero-point energy R_0 which, in molecules, corresponds to the core-core repulsion;

(ii) the element H_{jk} of the matrix H representing the one-particle part of $\hat{H}_{el}$ (the core Hamiltonian or the effective SCF Hamiltonian as the case may be);

(iii) the two-electron integrals representing the electrostatic energy of two 'electron clouds' having densities $\rho(\vec{x}) = \langle j|\hat{\delta}(\vec{r}-\vec{x})|k\rangle$, where $\hat{\delta}(\vec{r}-\vec{x})$ is the Dirac function for the position operator $\hat{\vec{r}}$.

The important point in (2.1.14) is that the choice of the basis orbitals is reflected explicitly in the Hamiltonian operator; indeed, any approximation made on the one-electron basis is automatically translated into $\hat{H}_{el}$ through the matrix elements of the one- and two-particle parts.

2.1.2. Many-electron basis: implications of limited CI. - Truncation of the M.E.S. basis (in view of the so-called limited CI) is not included explicitly in Eq. (2.1.14); it must be expressed by the additional condition that the matrix element of $\hat{H}_{el}$ (the 'coupling') between any M.E.S. belonging to the basis adopted and a M.E.S. not included in it must vanish. This condition gives rise to a number of relationships between the parameters appearing in Eq. (2.1.14): a number that can be very large when only a few M.E.S. are retained in the many-electron basis. For instance, with three orbitals and four electrons it is possible to construct fifteen M.E.S.; if we wish to keep only the five M.E.S. which correspond to single excitations from the second orbital, we have fifty conditions for couplings to vanish, which is close to the total number of variable parameters. Of course, those couplings may be negligible because of special circumstances; but they cannot be just *assumed* to vanish as if this implied minor consequences on the parameters and hence on the basis orbitals. We have thus another example of the way in which a method may become incoherent because of an assumption which is not legitimate. In the current literature this difficulty is often hinted at when justifications for neglect of certain configurations in CI calculations are sought; however, as far as we know, no systematic discussion of the general problem has been published so far.

2.2 Model Hamiltonians

Equation (2.1.14) is the fundamental expression of the present study. In accordance with our criterion for a legitimate approximate method, one should consider acceptable only a method to which a well defined Hamiltonian corresponds and whose structure does not imply contradictions with explicit or implicit assumptions.

A number of inconsistencies may be removed just by writing a suitable Hamiltonian, even if more of them may remain hidden in the parameters. (A well known example is the use of overlap to evaluate parameters associated with atomic orbitals otherwise assumed to form an orthonormal set.) In this section we shall give a few basic examples of model Hamiltonians and their applications.

As has been mentioned, the second quantization formalism is particularly convenient for models where a set of one-electron reference states - e.g. an atomic-orbital basis - is given. That basis is still assumed to be orthonormal; a restriction that will be discussed in the chapter devoted to the choice of the atomic orbital basis.

2.2.1. Non-interacting electrons. - The simplest example of a model Hamiltonian is the Hamiltonian of a system of non-interacting electrons, i.e. one which describes the electrons of a molecule in terms of an Independent-Particle Model (IPM). As the electron-interaction terms are zero, the Hamiltonian in question may be written in the form

$$\hat{H}_{IPM} = R_0 + \sum_{\mu,\nu,s} \langle \mu | \hat{H}_1 | \nu \rangle \, \hat{a}^+_{\mu s} \hat{a}_{\nu s} \qquad (2.2.1)$$

where the Greek suffixes μ, ν refer to the elements of a given atomic orbital basis.

The energy of a many-electron state of the basis (one that is represented by a Slater determinant over basis orbitals) is

$$\langle n_{1\uparrow} n_{1\downarrow} n_{2\uparrow} n_{2\downarrow} \ldots | \hat{H}_{IPM} | n_{1\uparrow} n_{1\downarrow} n_{2\uparrow} n_{2\downarrow} \ldots \rangle = R_0 + \sum_{\mu,s} \alpha_\mu n_{\mu s} \qquad (2.2.2)$$

where α_μ denotes the μ-th diagonal element of the matrix representing $\hat{H}_{IPM}$ over the basis AO's. No contribution is given to (2.2.2) by the "hopping" or "bond" parameters

$$\beta_{\mu\nu} = <\mu|\hat{H}_1|\nu> \qquad\qquad \mu \neq \nu$$

because the action of $\hat{a}^+_{\mu s}\hat{a}_{\nu s}$ on a state makes it either vanish or change to some other state; multiplication on the left by that state necessarily gives zero.

The coupling between two states of the many electron basis, associated with (2.2.1) is

$$<n_{1\uparrow}n_{1\downarrow}n_{2\uparrow}n_{2\downarrow}\ldots|\hat{H}_{IPM}|\bar{n}_{1\uparrow}\bar{n}_{1\downarrow}\bar{n}_{2\uparrow}\bar{n}_{2\downarrow}\ldots> = \beta_{\rho\tau}\,\delta(\rho s_\rho, \tau s_\tau) \qquad (2.2.3)$$

where $|\tau s_\tau>$ is a full spin orbital in the right-hand state and an empty spin orbital in the left-hand state, $|\rho s_\rho>$ the other way round; $\delta(\rho s_\rho, \tau s_\tau)$ is one if all the n's different from $n_{\tau s_\tau}$ and $n_{\rho s_\rho}$, coincide with the corresponding $\bar{n}$'s, and is zero otherwise.

Equation (2.2.3) means that the matrix elements of $\hat{H}_{IPM}$ between different basis states are either equal to zero or equal to one of the $\beta_{\mu\nu}$'s; the latter case is given when the right-hand state is generated from the left-hand one by the jump of an electron from the ν-th orbital to the μ-th orbital without any spin inversion.

To obtain the stationary states associated with $\hat{H}_{IPM}$ it is necessary, in principle, to diagonalize the matrix whose diagonal elements are given by (2.2.2) and whose off-diagonal elements are given by (2.2.3) over all the M.E.S. that can be obtained with the given set of atomic orbitals (singly-excited configurations, doubly-excited configurations, etc.). However, in the present case the problem can be solved more easily by taking into account that (a) the form (2.2.1) of the Hamiltonian $\hat{H}_{IPM}$ is valid for any set of basis orbitals, (b) according to (2.2.3) the matrix representing $\hat{H}_{IPM}$ over the basis M.E.S. will be diagonal if the hopping parameters associated with the basis *orbitals* vanish. These two remarks reduce to requiring that, given a set of atomic orbitals, we replace them by those linear combinations which make the off-diagonal elements of $\hat{H}_1$ vanish. In other words, if

$$|j> = \sum_\mu |\mu><\mu|j>, \quad \hat{a}^+_{\mu s} = \sum_j <j|\mu>\hat{a}^+_{js}, \quad \hat{a}_{\mu s} = \sum_k <\mu|k>\hat{a}_{ks}$$

$$\hat{n}_{\mu s} = \sum_{j,k} <j|\mu><\mu|k>\hat{a}^+_{js}\hat{a}_{ks}, \qquad (2.2.4)$$

then

$$<k|\hat{H}_1|j> = \sum_{\mu,\nu} <k|\nu><\nu|\hat{H}_1|\mu><\mu|j> = \sum_{\mu,\nu} <k|\nu>\beta_{\mu\nu}<\mu|j> \quad (\beta_{\mu\mu} \equiv \alpha_\mu) \tag{2.2.5}$$

and the latter expression vanishes for $k \neq j$, if the $|j>$ states diagonalize the matrix representation of $\hat{H}_1$. Then, by (2.2.3) the energy of a stationary state is just the energy of a M.E.S. $|\Phi>$ of the basis obtained from orbitals (2.2.4); by inserting in (2.2.3) k and $<k|\hat{H}|k>$ instead of μ and α_μ and using (2.2.5) we have

$$E = <\Phi|\hat{H}_{IPM}|\Phi> = R_0 + \sum_{k,s} n_{ks} <k|\hat{H}_1|k> = R_0 + \sum_{\mu}\left(\sum_{k,s} |<k|\mu>|^2 n_{ks}\right)\alpha_\mu + \sum_{\substack{\mu,\nu \\ \mu\neq\nu}}\left(\sum_{k,s} n_{ks} \frac{<k|\mu><\nu|k> + <k|\nu><\mu|k>}{2}\right)\beta_{\mu\nu} \tag{2.2.6}$$

where it has been taken into account that the AO's can be assumed to be real, and $\hat{H}_1$ is hermitean, so that $\beta_{\mu\nu} = \beta_{\nu\mu}$. Equation (2.2.6) is the well-known formula used in the Hückel method (and in EHT)

$$E = R_0 + \sum_{\mu} p_{\mu\mu}\alpha_\nu + \sum_{\mu,\nu}{}' p_{\mu\nu}\beta_{\mu\nu} \tag{2.2.7}$$

where the prime stands for $\mu \neq \nu$. Equation (2.2.7) is actually completely general; it holds if the basis orbitals $|k>$ do not diagonalize $\hat{H}_1$ and even if the state whose energy is to be evaluated is not just one basis MES. Then, the quantities $p_{\mu\mu}$ and $p_{\mu\nu}$ appearing in Eq. (2.2.7) are evidently defined as:

$$\begin{aligned} p_{\mu\mu} &= <\hat{n}_{\mu\uparrow}> + <\hat{n}_{\mu\downarrow}> \equiv n_{\mu\uparrow} + n_{\mu\downarrow} \\ p_{\mu\nu} &= \mathrm{Re}(<\hat{a}^+_{\mu\uparrow}\hat{a}_{\nu\uparrow}> + <\hat{a}^+_{\mu\downarrow}\hat{a}_{\nu\downarrow}>) \end{aligned} \tag{2.2.8}$$

and are the well-known elements of the "one-particle density matrix" i.e. of the "charge and bond-order" matrix (the signs < > denote the expectation value $<\Phi|\ |\Phi>$ over the given M.E.S.). In the specific case of a single M.E.S. over LCAO-MO's the above definition coincides with the definition derived from (2.2.6).

2.2.2. Intermediate models. - The independent particle model, in its strictest version, has a Hamiltonian that is obtained from the general one by formally neglecting the two-electron terms. There are other models intermediate between the IPM model and the full treatment. An example is the well known Hubbard model (1955a, b; 1957, 1958), whose intuitive foundation is the idea that electron-electron interactions just amount to standard modification of the hopping parameters *except* for electrons occupying the same atomic orbital. Therefore, Hubbard's model Hamiltonian in the AO basis is

$$\hat{H}_{Hu} = R_0 + \sum_{\mu,s} \alpha_\mu \hat{n}_{\mu s} + \sum_{\mu,\nu}' \sum_{s} \beta_{\mu\nu} \hat{a}^+_{\mu s} \hat{a}_{\nu s} + \sum_{\mu} (\mu\mu|\mu\mu) \hat{n}_{\mu\uparrow} \hat{n}_{\mu\downarrow} \tag{2.2.9}$$

where, as before, $\uparrow$ and $\downarrow$ denote opposite spins; account has been taken of the fact that, by the commutation rules (2.18a,b,c), $\hat{a}^+_{\mu\uparrow} \hat{a}^+_{\mu\downarrow} \hat{a}_{\mu\downarrow} \hat{a}_{\mu\uparrow} = \hat{a}^+_{\mu\uparrow} \hat{a}_{\mu\uparrow} \hat{a}^+_{\mu\downarrow} \hat{a}_{\mu\downarrow} \equiv$ $\equiv \hat{n}_{\mu\uparrow} \hat{n}_{\mu\downarrow}$. This time the energy of a basis state is given by

$$\begin{aligned} \langle n_{1\uparrow} n_{1\downarrow} n_{2\uparrow} n_{2\downarrow} \ldots | \hat{H}_{Hu} | n_{1\uparrow} n_{1\downarrow} n_{2\uparrow} n_{2\downarrow} \ldots \rangle &= R_0 + \\ &+ \sum_{\mu,s} \alpha_\mu n_{\mu s} + \sum_{\mu} (\mu\mu|\mu\mu) n_{\mu\uparrow} n_{\mu\downarrow} \; . \end{aligned} \tag{2.2.10}$$

If the state in question is built over molecular orbitals $|j\rangle$, $|k\rangle, \ldots,$ with $n_{\mu\uparrow} = n_{\mu\downarrow}$, the corresponding energy is (cf. Eq. (2.2.7))

$$E = R_0 + \sum_{\mu} P_{\mu\mu} (\alpha_\mu + \frac{1}{4} P_{\mu\mu} J_{\mu\mu}) + \sum_{\mu,\nu}' \beta_{\mu\nu} P_{\mu\nu} \tag{2.2.11}$$

where $J_{\mu\mu}$ stands for $(\mu\mu|\mu\mu)$. Different many electron states (M.E.S.) over the molecular orbitals that diagonalize the one-electron part of $\hat{H}_{Hu}$ do not necessarily give a vanishing matrix element, for

$$\begin{aligned} \sum_{\mu} (\mu\mu|\mu\mu) \hat{n}_{\mu\uparrow} \hat{n}_{\mu\downarrow} &= \sum_{\mu} \sum_{j,k} \sum_{l,m} \langle j|\mu\rangle \langle \mu|k\rangle \langle l|\mu\rangle \cdot \\ &\cdot \langle \mu|m\rangle (\mu\mu|\mu\mu) \hat{a}^+_{j\uparrow} \hat{a}_{k\uparrow} \hat{a}^+_{l\downarrow} \hat{a}_{m\downarrow} \end{aligned} \tag{2.2.12}$$

so that a contribution is provided to an off-diagonal element by all those pairs of M.E.S. whose partners can be made equal either by shifting one electron from the m-th orbital (with spin $\downarrow$) to the l-th one (spin $\uparrow$), and/or by shifting one electron from the k-th orbital (spin $\uparrow$) to the j-th orbital (spin $\uparrow$).

An example of application of the Hubbard Hamiltonian can be found in a study of small clusters of Be by Torrini et al. (1976).

2.2.3. <u>The pairing theorem.</u> - To continue in this general illustration of model Hamiltonians in the second-quantization formalism, we introduce now the 'pairing theorem' of McLachlan. That theorem applies in the frame of the so-called P-P-P (Pariser-Parr-Pople) method for π electrons, a method with which a well-defined model Hamiltonian can be associated (Koutecky 1967; cf. sec. 4.5):

$$\hat{H}_{PPP} = R_0 + \hat{H}^{(\uparrow)}_{PPP} + \hat{H}^{(\downarrow)}_{PPP}$$

where (with $s = \uparrow, \downarrow$)

$$\hat{H}^{(s)}_{PPP} = \sum_{\mu} \alpha_{\mu} \hat{n}_{\mu s} + \sum_{\mu,\nu}{}' \beta_{\mu\nu} \hat{a}^{+}_{\mu s} \hat{a}_{\nu s} + \frac{1}{2} \sum_{\mu,\nu,s} (\mu\mu|\nu\nu) \hat{a}^{+}_{\mu s} \hat{a}^{+}_{\nu s'} \hat{a}_{\nu s'} \hat{a}_{\mu s} \qquad (2.2.13)$$

$\beta_{\mu\nu}$ being zero for non-linked atoms.

The basis to which the subscript μ, ν refer consists of one π orbital per atom - like the $2p\pi$ orbital of the benzene carbon atoms. Suppose now that the given π system is alternant (Coulson and Rushbrooke 1940), i.e. that we can divide the atoms carrying π atomic orbitals into two sets, such that every atom of the first set is linked only to atoms of the second set, and vice versa. (According to a current notation, the atoms of one set will be distinguished by a star, and called 'starred atoms'; the others will be the 'unstarred atoms'. Scheme I illustrates that notation.)

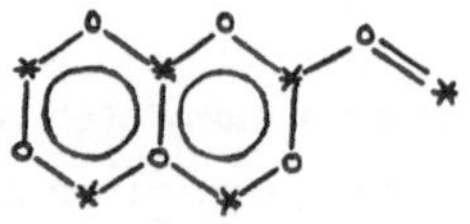

I

Next, we change the basis by changing the signs of the AO's of the starred atoms (and leaving the other AO's unchanged). Equation (2.2.13) becomes

$$\hat{H}^{(s)}_{PPP} = \sum_{\mu} \alpha_{\mu} \hat{n}_{\tilde{\mu} s} - \sum_{\tilde{\mu},\tilde{\nu}} \sum \beta_{\mu\nu} \hat{a}^{+}_{\tilde{\mu} s} \hat{a}_{\tilde{\nu} s} + \frac{1}{2} \sum_{\tilde{\mu},\tilde{\nu},s} (\mu\mu|\nu\nu) \hat{a}^{+}_{\tilde{\mu} s} \hat{a}^{+}_{\tilde{\nu} s'} \hat{a}_{\tilde{\nu} s'} \hat{a}_{\tilde{\mu} s} \qquad (2.2.14)$$

where the bar over the subscripts refers to the new basis.

Finally, we introduce the notation

$$\hat{\omega}^{+}_{\mu s} = \hat{a}_{\tilde{\mu} s} \qquad \hat{\omega}_{\mu s} = \hat{a}^{+}_{\tilde{\mu} s} \qquad (2.2.15)$$

which means that the annihilation and creation operators over the new basis are interpreted as hole creation and hole annihilation operators, respectively. Remembering that Eq.(2.1.8) hold for any orthogonal basis, we have

$$\hat{\omega}^{+}_{\mu s}\hat{\omega}_{\nu s} = \delta_{\mu\nu} - \hat{a}^{+}_{\underset{\sim}{\mu s}}\hat{a}_{\underset{\sim}{\nu s}} \tag{2.2.16}$$

whence

$$\begin{aligned}\hat{H}^{(s)}_{PPP} &= \sum_{\mu}\alpha_{\mu}(1-\hat{\omega}^{+}_{\mu s}\hat{\omega}_{\mu s}) + \sum_{\mu,\nu}{}' \beta_{\mu\nu}\hat{\omega}^{+}_{\mu s}\hat{\omega}_{\nu s} + \\ &+ \frac{1}{2}\sum_{\mu,\nu}(\mu\mu|\nu\nu) - \frac{1}{2}\sum_{\mu,\nu}(\mu\mu|\nu\nu)\hat{\omega}^{+}_{\mu s}\hat{\omega}_{\mu s} + \\ &- \frac{1}{2}\sum_{\mu,\nu}(\mu\mu|\nu\nu)\hat{\omega}^{+}_{\nu s}\hat{\omega}_{\nu s} + \frac{1}{2}\sum_{\mu,\nu}(\mu\mu|\nu\nu)\hat{\omega}^{+}_{\mu s}\omega^{+}_{\nu s'}\omega_{\nu s'}\hat{\omega}_{\mu s}.\end{aligned} \tag{2.2.17}$$

If $\sum_{\mu}\alpha_{\mu}\hat{\omega}^{+}_{\mu s}\hat{\omega}_{\mu s}$ is added and subtracted, the following expression is obtained:

$$\begin{aligned}\hat{H}^{(s)}_{PPP} &= \sum_{\mu}\alpha_{\mu}\hat{\omega}^{+}_{\mu s}\hat{\omega}_{\mu s} + \sum_{\mu,\nu}{}'\beta_{\mu\nu}\hat{\omega}^{+}_{\mu s}\hat{\omega}_{\nu s} + \frac{1}{2}\sum_{\mu,\nu}(\mu\mu|\nu\nu)\hat{\omega}^{+}_{\mu s}\hat{\omega}^{+}_{\nu s'}\hat{\omega}_{\nu s'}\hat{\omega}_{\mu s} + \\ &+ \sum_{\mu}\left[\alpha_{\mu} + \frac{1}{2}\sum_{\nu}(\mu\mu|\nu\nu)\right](1 - 2\hat{\omega}^{+}_{\mu s}\hat{\omega}_{\mu s}) .\end{aligned} \tag{2.2.18}$$

Equation (2.2.18) means that for a given number of 'holes' we shall have the same result as for the same number of 'particles' (i.e. when the same number of spin orbitals of the total spin orbital basis is either empty or full) except for the last term. But suppose we are considering the energy of a given state; then the operator product in the last term just multiplies the given state by 1 if the μs-th spin orbital is empty, by 0 if it is occupied. Therefore, the contribution to the total energy given by the term in question is

$$\Delta E = \sum_{\mu s}\left[\alpha_{\mu} + \frac{1}{2}\sum_{\nu}(\mu\mu|\nu\nu)\right] - 2\sum_{\mu s\ \text{empty}}\left[\alpha_{\mu} + \frac{1}{2}\sum_{\nu}(\mu\mu|\nu\nu)\right] . \tag{2.2.19}$$

We now proceed to remark that the PPP parametrization assigns a standard value U to the expression in square brackets of the last equation, if it refers to a carbon atom. Then (2.2.19) becomes

$$\Delta E = UM - 2U(M - N) = U(2N - M) = 2UQ \quad \text{with} \quad Q = N - \frac{1}{2}M \tag{2.2.20}$$

where account has been taken of the fact that the number of holes is M - N if M is the total number of spin orbitals and N the number of electrons - and Q is evidently the electronic charge of the molecule, for N = M means that there are precisely the valence electrons of the given carbon atoms.

In conclusion, we can state the theorem that in the case of a π hydrocarbon system with M π orbitals: (i) the PPP Hamiltonian is the same up to a constant 2MU for holes as for electrons (except that in the former case α_μ must be replaced by $-\alpha_\mu$) - in particular, the states of the system having N electrons are in a one-to-one correspondence with those of the system having N holes, i.e. M - N electrons; (ii) the PPP energy of an N electron system differs from that of the 'paired' system having M - N electrons by 2UQ - the expectation value of the operator that makes (2.1.32) different from the Hamiltonian of a system of N holes (McLachlan 1959).

2.3 Matrix Formalism. Inclusion of Overlap

The above examples are intended to provide a concrete illustration of model Hamiltonians in the second quantization formalism. We shall use the ideas presented here in a more systematic form in discussing the various semiempirical methods.

As should be clear from the foregoing discussion, the matrix formalism currently used in quantum chemistry to translate methods into computer programs is also a standard part of the second-quantization scheme. Therefore, we close the present introductory chapter by presenting our notation concerning matrices; we shall also take the opportunity to introduce overlap explicitly.

2.3.1. Orbital bases and one-electron operators. - The basis will be represented as a row-matrix. Let us concentrate for the moment on the one-electron case. The AO basis will be denoted by

$$|\mu\rangle \equiv (|1\rangle_G \ldots |\mu\rangle\, |\nu\rangle \ldots) \qquad (2.3.1a)$$

and an equivalent MO basis by

$$|k\rangle \equiv (1\rangle_L\, |2\rangle_L \ldots |j\rangle\, |k\rangle \ldots) \qquad (2.3.1b)$$

the subscripts G and L being used only when it is impossible to distinguish elements of either basis otherwise. The corresponding spin orbital bases will be written

$$|\mu, s\rangle \equiv |\mu_\uparrow\rangle |\mu_\downarrow\rangle; \qquad |k, s\rangle \equiv |k_\uparrow\rangle |k_\downarrow\rangle \tag{2.3.1c}$$

and the subscripts $\downarrow$ and $\uparrow$ will be dropped if - as is often the case - the orbitals used to build spin orbitals of spin $\uparrow$ are the same as those used for spin $\downarrow$ - which is a current assumption, but not a necessary one (de Heer, 1963).

The "bra" bases associated with those given above will be represented by *column* matrices, whose elements are the bras of the given orbitals.

We must now associate to the bases (2.3.1) and (2.3.2) two sets of mathematical quantities, creation and annihilation operators and matrices proper (i.e., matrices with numerical elements). The matrices represent either the standard first-quantization operators or linear transformations.

The former can be thought of as formally given by products of the type

$$\langle \mu|\hat{Q}|\mu\rangle \equiv \begin{vmatrix} {}_G\langle 1| \\ {}_G\langle 2| \\ \vdots \\ \langle \mu| \\ \vdots \\ \langle \nu| \\ \vdots \\ {}_G\langle M| \end{vmatrix} \left(\hat{Q}|1\rangle_G \; \hat{Q}|2\rangle_G \; \ldots \; \hat{Q}|\mu\rangle \; \ldots \; \hat{Q}|\nu\rangle \; \ldots \; \hat{Q}|M\rangle_G \right)$$

Thus, in particular, the matrix representing the unit operator $\hat{I}$ - i.e. the metric (overlap) matrix associated with our basis - is given by:

$$\text{(a)} \quad S^{(G)} \equiv \langle \mu|\mu\rangle \quad \text{or} \quad S_{\mu\nu} = \langle \mu|\nu\rangle , \tag{2.3.3a}$$

$$\text{(b)} \quad S^{(L)} \equiv \langle k|k\rangle \quad \text{or} \quad S_{jk} = \langle j|k\rangle , \tag{2.3.3b}$$

$$\text{(c)} \quad S^{(G,s)} \equiv \langle \mu, s|\mu, s\rangle = \begin{vmatrix} S^{(G)} & 0 \\ 0 & S^{(G)} \end{vmatrix} , \tag{2.3.3c}$$

where the superscript (G,s) refers to the basis (2.2.1c) and the block diagonal form of $S^{(G,s)}$ is a consequence of the orthogonality of states with different spins (the two blocks refer to spin $\uparrow$ and spin $\downarrow$, respectively). The matrix representing the one-electron Hamiltonian operator $\hat{H}_1$ is

(a) $H^{(G)} \equiv \langle \mu|\hat{H}_1|\mu\rangle$ (2.3.4a)

(b) $H^{(L)} \equiv \langle k|\hat{H}_1|k\rangle$ (2.3.4b)

(c) $H^{(G,s)} \equiv \left|\begin{array}{c|c} H^{(G)} & 0 \\ \hline 0 & H^{(G)} \end{array}\right| .$ (2.3.4c)

2.3.2. Linear transformations. - Linear transformations are represented by matrices whose columns correspond to the elements of the new set, and whose rows correspond to the old basis set. Transformations carrying from one basis to an equivalent one are square non-singular matrices. For instance,

$$|k\rangle = |\mu\rangle C \qquad \langle k| = C^+\langle\mu| \tag{2.3.5}$$

and, if the old basis had a metric matrix S, and the new one has a metric matrix S', we must have

$$S' = \langle k, s|k, s\rangle = C^+SC \tag{2.3.6a}$$

which, if the basis k is an orthonormal one, reduces to

$$C^+SC = I \tag{2.3.6b}$$

with I the unit matrix; thus C must be a square non-singular matrix.

The transformation T that carries from a non-orthogonal basis $|\mu\rangle_{AO}$ with metric matrix S to the nearest orthonormal basis $|\mu\rangle_{LOAO}$ has been shown by Löwdin (1950) to be

$$T = S^{-1/2}, \quad \text{i.e.} \quad |\mu\rangle_{LOAO} = |\mu\rangle_{AO}S^{-1/2} . \tag{2.3.7}$$

We can identify the orthogonal AO's used so far with LOAO's, and the problem of introducing overlap reduces to the problem of passing to the basis $|\mu\rangle_{AO}$.

By analogy with (2.3.6a) it is seen immediately that

$$H = \langle u'|\hat{H}_1|\mu'\rangle = T\langle\mu|\hat{H}_1|\mu\rangle T = S^{-1/2}H^{(AO)}S^{-1/2} \tag{2.3.8}$$

where primed orbitals are LOAO's and unprimed ones are AO's. Moreover, the two-electron integrals can be expressed by introducing twice the above transformation:

$$(\mu'\nu'|\rho'\sigma')_{LOAO} = \sum_{\substack{\mu,\nu \\ \rho,\sigma}} T_{\mu'\mu} T_{\nu'\nu} T_{\rho'\rho} T_{\sigma'\sigma} < \mu\nu|\rho\sigma > \tag{2.3.9}$$

where account has been taken of the fact that T is a hermitean matrix, because such is S.

2.3.3. Creation-annihilation operators and the Hamiltonian. - Keeping in mind that the overall expression of an operator must be independent of the basis, we can now write the expression of the Hamiltonian operator over non-orthogonal basis orbitals.

First of all, we introduce the operator matrices

$$\hat{a}_s^+ = (\hat{a}_{1s}^+ \hat{a}_{2s}^+ \hat{a}_{3s}^+ \hat{a}_{4s}^+ \dots) \qquad s = \uparrow, \downarrow \tag{2.3.10}$$

$$\hat{a}_s = \begin{pmatrix} \hat{a}_{1s} \\ \hat{a}_{2s} \\ \hat{a}_{3s} \\ \hat{a}_{4s} \\ \vdots \end{pmatrix} \qquad s = \uparrow, \downarrow \tag{2.3.11}$$

and the supermatrix

$$(W)_{\mu\nu\rho\tau} = (W_{\mu\nu})_{\rho\tau} = W_{\mu\rho\tau\nu} \; .$$

Therefore, we can write in an orthogonal basis, say $|\mu>_{LOAO}$

$$\hat{H} = R_0 + \sum_s \hat{a}^+ H \hat{a}_s + \frac{1}{2} \sum_{s,s'} \hat{a}_s^+ (\hat{a}_{s'}^+, W\hat{a}_{s'}) \hat{a}_s \tag{2.3.12}$$

where H is the matrix of the one-electron part of the Hamiltonian over the given basis. Next let us call $\hat{b}^+$ and $\hat{b}$ the creation-annihilation operator matrices corresponding to a non-orthogonal basis, say $|\mu>_{AO}$. We assume that the relations

$$\hat{b}_s = X\hat{a}_s \qquad \text{and} \qquad \hat{b}_s^+ = \hat{a}_s^+ X^+$$

hold, with X a linear transformation to be determined. Then

$$\hat{a}_s^+ H \hat{a}_s = \hat{b}_s^+ (X^+)^{-1} H \, X^{-1} \, \hat{b}_s$$

and since, in the basis $|\mu\rangle_{LOAO}$, Eq. (2.3.8) holds, we shall have

$$\hat{a}_s^+ H \hat{a}_s = \hat{b}_s^+ H^{(AO)} \hat{b}_s$$

if

$$X = S^{-1/2} \equiv T \, ; \quad \hat{b}_s = T\hat{a}_s \, , \quad \hat{b}_s = \hat{a}_s^+ T \; . \tag{2.3.13}$$

If can be verified that (2.3.13) also gives an invariant two-electron term if the two-electron integrals $W_{\mu\rho\tau\nu} \equiv (\mu\nu|\rho\tau)$ over $|\mu\rangle_{LOAO}$ are replaced by those over $|\mu\rangle_{AO}$. In conclusion, the expression (2.3.12) is also valid in a non-orthogonal basis:

$$\hat{H} = R_0 + \Sigma \; H_{\mu\mu} \hat{b}^+_{\mu s} \hat{b}_{\nu s} + \frac{1}{2} \Sigma \; W_{\mu\rho\tau\nu} \hat{b}^+_{\mu s} \hat{b}^+_{\rho s'} \hat{b}_{\tau s'} \hat{b}_{\nu s} \; . \tag{2.3.14}$$

The only points to be kept in mind are that (i) the anti-commutation relations (2.1.8) now become

$$\hat{b}^+_{\mu s} \hat{b}^+_{\nu s} + \hat{b}^+_{\nu s} \hat{b}^+_{\mu s} = 0 \tag{2.3.15a}$$

$$\hat{b}_{\mu s} \hat{b}_{\nu s} + \hat{b}_{\nu s} \hat{b}_{\mu s} = 0 \tag{2.3.15b}$$

$$\hat{b}^+_{\mu s} \hat{b}_{\nu s} + \hat{b}_{\nu s} \hat{b}^+_{\mu s} = (S^{-1})_{\mu\nu} \tag{2.3.15c}$$

and (ii) the operators $\hat{b}_{\mu s}$ and $\hat{b}^+_{\mu s}$ annihilate and create electrons in the orbitals of the 'dual' basis (Grimley 1970):

$$|\bar{\mu}\rangle_{AO} = |\mu\rangle_{AO} S^{-1} \tag{2.3.15d}$$

rather than in the basis $|\mu\rangle_{AO}$ itself.

2.3.4. <u>Eigenvalue equation for $\hat{H}$.</u> - In the matrix formalism, the eigenvalue equation for the Hamiltonian operator - currently called the Schrödinger equation for stationary states - is written

$$HC = SCE \; . \tag{2.3.16}$$

This equation, as is well known, defines the stationary state energies and the corresponding components of the eigenstates over the given basis (columns of C). It is accompanied by the normalization condition (2.2.6b). It is useful to remember that (2.3.16) also defines a basis in the given Hilbert space, the basis in which H is diagonal (cf. sec. 2.3.2). Therefore,

$$(C)_{\mu k} \equiv \langle \bar{\mu}|k\rangle, \quad \text{and} \quad (C^{+})_{k\mu} \equiv \langle k|\bar{\mu}\rangle \tag{2.3.17}$$

where the dual basis (2.3.15d) appears because the identity

$$E = \langle k|\hat{H}|k\rangle = \langle k|\bar{\mu}\rangle\langle\mu|\hat{H}|\mu\rangle\langle\bar{\mu}|k\rangle = C^{+}HC \tag{2.3.18}$$

cannot hold if $|\bar{\mu}\rangle\langle\mu|$ is replaced by $|\mu\rangle\langle\mu|$.

Equation (2.3.16) is the matrix equation for the overall many-electron problem if the basis is one of M.E.S. and is given by (2.3.14) or (2.3.12); it is the equation for the molecular orbitals in an LCAO basis (Slater type AO's, Gaussian AO's, etc.) and $\hat{H}$ represents an effective (semi-empirical or ab-initio, possibly SCF) one-electron Hamiltonian.

Equation (2.2.16) is over *finite* matrices if the range of variation of the basis 'subscripts' is finite; it is over *discrete* matrices if the 'subscripts' characterizing the various basis elements form a denumerable set.

The various methods of quantum chemistry normally involve finite discrete matrices - i.e. truncated discrete bases -. This limitation makes all methods quantitatively debatable, even when they give reasonable or even excellent energy values. In general, the choice of the basis is a critical point of all methods and models in the physics of molecules and crystals. We shall devote a separate chapter to it.

2.4. Spectral Decomposition of the Hamiltonian and Effective Potentials

The second quantization formalism is especially useful when a one-particle basis is given, and the Hamiltonian is known in terms of one-electron and two-electron matrix elements over that basis. A more general formalism has been taken as a starting point by Durand (1976) in order to elaborate a method for deriving effective Hamiltonians corresponding to simple one- and two-electron interactions.

2.4.1. Spectral resolution and projection operators. - If the eigenvalues E_i and the corresponding eigenvectors $|\Psi_i\rangle$ of a Hamiltonian operator $\hat{H}$ are given, it is easy to reconstruct $\hat{H}$ by writing its spectral resolution

$$\hat{H} = \sum_i E_i \hat{P}_i \qquad \left(\text{with} \sum_i P_i = \hat{1}\right) \tag{2.4.1}$$

(to be modified by the inclusion of integrals if the spectrum of $\hat{H}$ is not completely discrete). Here, $\hat{1}$ is the identity operator, and

$$\hat{P}_i = |\Psi_i\rangle\langle\Psi_i|$$

is the projection operator associated with the i-th eigenstate: by acting on any state vector $|A\rangle$ it extracts from it the contribution of $|\Psi_i\rangle$:

$$\hat{P}_i|A\rangle \equiv |\Psi_i\rangle\langle\Psi_i|A\rangle . \tag{2.4.2}$$

Suppose now that we are interested in a set of eigenstates of $\hat{H}$ which span a given subspace Ω of the Hilbert space of $\hat{H}$, in particular an eigensubspace of $\hat{H}$, i.e. one whose elements produce again elements of Ω when subjected to $\hat{H}$. Then, we can define two projection operators

$$\hat{P} = \sum_{i\in\Omega} \hat{P}_i , \qquad \text{and} \qquad \hat{Q} = \sum_{i\notin\Omega} \hat{P}_i , \tag{2.4.3}$$

which add up to the unit operator. Using $\hat{P}$ and $\hat{Q}$ we can split $\hat{H}$ into two operators one of which coincides with $\hat{H}$ in Ω and vanishes outside Ω, the other which has exactly the opposite property. In fact, noting that

$$\hat{P}\hat{H}\hat{Q} = \hat{Q}\hat{H}\hat{P} = 0 \qquad \text{if } \Omega \text{ is an eigenspace of } \hat{H} , \tag{2.4.3a}$$

we have

$$\hat{H} = \hat{1}\cdot\hat{H}\cdot\hat{1} = \hat{P}\hat{H}\hat{P} + \hat{Q}\hat{H}\hat{Q} = \sum_{j\in\Omega} E_j\hat{P}_j + \sum_{j\notin\Omega} E_j\hat{P}_j . \tag{2.4.4}$$

A more general operator vanishing outside Ω and having in Ω the same eigenvalues as $\hat{H}$, but not necessarily the same eigenvectors, can also be defined in terms of its spectral resolution and of projection operators $\hat{P}_i'$, associated with a basis of states $|\Phi_i\rangle$ in Ω:

$$\hat{H}_\Omega = \sum_{i\in\Omega} E_i\hat{P}_i' = \hat{P}'\hat{H}\hat{P}' = \sum_{i\notin\Omega} \hat{U}|\Psi_i\rangle E_i \langle\Psi_i|\hat{U}^+ = \hat{U}\hat{P}\hat{H}\hat{P}\hat{U}^+ \tag{2.4.5}$$

where $\hat{U}$ is defined as the unitary operator which takes $|\Psi_i\rangle$ into $|\Phi_i\rangle$ (that operator must exist because $|\Psi_i\rangle$ and $|\Phi_i\rangle$ both belong to Ω).

2.4.2. Durand's effective Hamiltonians. - Nicolas and Durand (1978) define an effective Hamiltonian as one which has the same eigenvalues as the exact one in a certain energy range (and in a certain Hilbert subspace Ω), and is arbitrary in the rest of the Hilbert space: they call 'variational' an effective Hamiltonian whose 'good' eigenvalues all lie below the spurious ones. From what has been reported in the preceding subsection, that Hamiltonian must coincide with an $\hat{H}_\Omega$ of the form (2.4.5) within Ω, and can be anything elsewhere (Ω being as before an eigenspace of the correct $\hat{H}$). This corresponds to the conditions:

$$\hat{P}\hat{H}\hat{P} = \hat{P}\hat{H}_{eff}\hat{P} , \tag{2.4.6}$$

$$\hat{P}\hat{H}_{eff}\hat{Q} = \hat{Q}\hat{H}_{eff}\hat{P} = 0 . \tag{2.4.7}$$

These conditions are equivalent to requiring that the norm of the difference between $\hat{H}$ and $\hat{H}_{eff}$ be zero:

$$(\hat{H} - \hat{H}_{eff}|\hat{H} - \hat{H}_{eff}) = 0 \tag{2.4.8}$$

(the scalar product of two operators $\hat{A}$ and $\hat{B}$ belonging to the vector space of the operators associated with the given Hilbert space is defined here as:

$$(\hat{A}|\hat{B}) = \mathrm{Tr}(\hat{P}\hat{A}^+\hat{P}\hat{B}\hat{P}) + \mathrm{Tr}(\hat{P}\hat{A}^+\hat{Q}\hat{B}\hat{P}) + \mathrm{Tr}(\hat{Q}\hat{A}^+\hat{P}\hat{B}\hat{Q}) , \tag{2.4.9}$$

the Hermiticity condition being fulfilled.)

Now, in practical applications an eigenspace of $\hat{H}$ is only approximately known - for instance, the π Slater determinants are assumed to span an eigensubspace of the Hilbert space of a conjugated molecule, but actually the eigenstates will often include some σ admixture. Therefore, Nicolas and Durand suggest that (2.4.8) should be replaced by

$$(\hat{H} - \hat{H}_{eff}|\hat{H} - \hat{H}_{eff}) = \min. \tag{2.4.10}$$

Expression (2.4.10) is an equation for $\hat{H}_{eff}$ which can be easily translated into a familiar matrix equation if the operator in question is defined in a basis of operators $\hat{A}_j$ - e.g. the products of creation-annihilation operators appearing in the second quantization expression of a Hamiltonian -:

$$\hat{H}_{eff} = \hat{R}_0 + \sum_j c_j\hat{A}_j , \tag{2.4.11}$$

$\hat{R}_0$ being a constant operator (e.g. the core repulsion). Then Eq.(2.4.10) reduces to an equation for the coefficients c_j:

$$c_j = (\hat{A}_j|\hat{H} - \hat{R}_0) . \tag{2.4.12}$$

Nicolas and Durand (1979) give the mathematical details and the derivation of various effective Hamiltonians from the above equations. We shall only consider the application to pseudopotential methods.

2.4.3. Effective Hamiltonians and Pseudopotentials. - The pseudopotential concept is introduced when one tries to treat only part of the electrons of an atom or molecule. For instance, in the case of n valence electrons an n-electron Hamiltonian $\hat{H}_{ps}$ is written where the ordinary Coulomb potential-energy operator between an electron and the atomic core M is replaced by an appropriate potential $W_{ps,M}$ such that the valence orbitals - or in general the orbitals whose energies fall within a certain range - have the same energies as would be predicted from the exact Hamiltonian. In Durand's view (1976) $W_{ps,M}$ must be a fixed and *transferable* one-electron operator determined once for all from selected molecules or atoms: the equality of eigenvalues will then be approximate.

The SCF version of $\hat{H}_{ps}$ is a one-electron operator $\hat{F}_{ps}$ which can be written in the form

$$\hat{F}_{ps} = \sum_{v} \epsilon_v |\Phi_v\rangle\langle\Phi_v| + \sum_{\epsilon} \epsilon_e |\Phi_e\rangle\langle\Phi_e| , \qquad (2.4.13)$$

where the Φ states are orbitals, the ϵ's are their orbital energies, the subscript v refers to the valence SCF orbitals, the subscript e to the other ('excited') SCF orbitals obtained from $\hat{F}_{ps}$; the first summation is to have the same e_v values as the rigorous SCF Hamiltonian $\hat{F}$. An effective many-electron Hamiltonian is then written in the second-quantized form over the orbitals associated with $\hat{F}_{ps}$:

$$\hat{H}_{eff} = R_0' + \sum \epsilon_v \hat{n}_v + \sum \epsilon_e \hat{n}_e , \qquad (2.4.14)$$

with

$$R_0' = R_0 - \sum [(vv|v'v') - (vv'|v'v)] \qquad (2.4.15)$$

R_0 being the nuclear repulsion and R_0' an effective core repulsion of the type discussed in sec. 4.2.2.

The Hamiltonian (2.4.14) may be cast into the form (2.4.11) by expanding the number operators over the SCF orbitals of $\hat{F}_{ps}$ into products of creation-annihilation operators over the spin orbitals of a standard AO basis; thus the technique described in Eqs. (2.4.10)-(2.4.12) can be used to determine the Hamiltonian (2.4.14) in matrix form; from the latter the pseudopotentials can finally be derived given a general form (Barthelat 1977). An example will be given for the Hückel method in sec. 3.3.4.

Chapter 3. One-Electron Schemes

The one-electron schemes of quantum chemistry can be divided into three classes: Hückel-type methods, semiempirical SCF models, ab-initio SCF model. We include in the second class also the so-called density-functional methods, even though they do not belong to it in a strict sense. We shall discuss first of all the models of the first class, then we shall derive the equations for the SCF methods, and discuss the physical meaning of those simpler models; then we shall give the SCF forms of the best known semiempirical schemes; we shall briefly consider the current *ab initio* SCF models only with reference to the hidden semiempirical features associated with the choice of the basis and of the geometry.

3.1. Hückel-Type Methods

The Hückel model Hamiltonian has been given in Section 2.1.3.2, where it was treated as a special case of the general IPM *sensu stricto* (i.e. not involving any SCF condition). We now proceed to a more specific discussion.

The model underlying the Hückel method is one where the field of all the electrons but one is constant and where the orbitals of the atoms span a linear subspace of the one-electron state space. The parameters of the model are the elements of the H matrix, they are supposed to be characteristic of atoms and bonds. A complete list of specifications of the historical Hückel method (Hückel 1930) is as follows:

Type of Hamiltonian: IPM non-SCF

Basis: minimal orthogonal or orthogonalized AO basis - one valence orbital per atom (normally a π one, but occasionally, as with metals, an s orbital).

Atomic parameters (diagonal elements of the Hamiltonian matrix over the given basis): standard values, possibly with corrections for inductive effects (Orgel *et al.* 1951), i.e. with 10 to 30 % of the atomic parameters of heteroatoms added to the atomic parameters of the carbon atoms directly bound to them.

Bond parameters (off-diagonal elements of the Hamiltonian matrix; also called resonance integrals): standard values β_{AB} between directly linked atoms AB, zero otherwise (nearest neighbour approximation: n.n.a.).

Typical lists of parameters are given in Table 3.1 (Del Re 1960b, Wentworth et al. 1975).

Table 3.1. Parameters for the Hückel Method (in the notation of Eq. (3.1.5).

A) Parameters in heterocycles adapted to π dipole moment calculations (Orgel *et al.* 1951)[1]

δ_C	$= 0.1\sum \delta_x$ [2]	η_{CC}	$= 1.0$
$\delta_{N(imino)}$	$= 1.0$	η_{CN}	$= 1.2$
$\delta_{N(amino)}$	$= 2.0$	η_{CO}	$= 1.4$
$\delta_{O=}$	$= 2.0$	η_{NO}	$= 0.6$
$\delta_{O<}$	$= 3.2$		

B) Parameters for substituents in benzenoid hydrocarbons from electron affinities (Wentworth *et al.* 1975)[3]

α	$= -1.42$ ev	β	$= -1.79$ ev
δ_C	$= 0.4\sum \delta_x$	δ_{Cl}	$= 2.90$
δ_O	$= 0.96$	η_{CO}	$= 1.37$
δ_F	$= 2.31$	η_{CF}	$= 1.47$
δ_{CH3}	$= -0.12$	η_{CCl}	$= 1.67$

[1]: Calculated dipole moments should be divided by 1.6 to match experimental data.
[2]: x denotes heteroatoms adjacent to C.
[3]: least-square adjustment on 21 aromatic aldehydes and ketones, plus 1-Chloronaphtalene.

Three questions may be discussed in connection with the Hückel method, as well as with other simple methods: (i) its model Hamiltonian formulation; (ii) the underlying physical picture; (ii) its range of applicability and the correct use of its results. The model formulation is quite straightforward:

$$\hat{H}_{Hü} = R_0 + H^{\uparrow}_{Hü} + H^{\downarrow}_{Hü} \tag{3.1.1a}$$

$$\hat{H}^{(s)}_{Hü} = \sum \alpha_\mu \hat{n}_{\mu s} + \sum{}' \beta_{\mu\nu} \hat{a}^+_{\mu s} \hat{a}_{\nu s} \tag{3.1.1b}$$

but is accompanied by conditions on the basis orbitals (the AO's of the Hückel basis $|\chi_{Hü}\rangle$) that are rather stringent. We can assume that an effective one-electron Hamiltonian $\hat{H}_{1Hü}$ is given in the configuration space of one electron, and thus:

$$\alpha_\mu = \langle\mu|\hat{H}_{1Hü}|\mu\rangle\ , \quad \beta_{\mu\nu} = \langle\mu|\hat{H}_{1Hü}|\nu\rangle\ , \quad \langle\mu|\nu\rangle = \delta_{\mu\nu}\ ,$$
$$\langle\mu|\hat{H}_{1Hü}|\mu'\rangle = 0\ , \quad \langle\mu|\mu'\rangle = 0 \tag{3.1.2}$$

where $|\mu'\rangle$ is any orbital of a set $|\chi'\rangle$ which can be added to the Hückel AO basis to complete it. This is equivalent to saying that the elements of the Hückel basis are assumed to span a linear subspace of the one-electron state-space, *i.e.* that the Hückel states are correctly described by the limited basis $|\chi\rangle_{Hü}$. Indeed, an assumption characteristic of the Hückel method in its original version for conjugated organic molecules is the so-called σ-π separation (Lykos and Parr 1956), which amounts precisely to claiming that certain states of conjugated molecules can be treated as a problem in a subspace of the complete one-electron state-space associated with the given molecule. We shall return to this condition in discussing the basis problem. Then we shall also consider the meaning of the term 'atomic orbitals' used in the present context: a very delicate problem, as will be seen.

As has been mentioned, the physical meaning of non-SCF Hamiltonians is best discussed in terms of SCF ones. However, it is easy to describe the physical picture which led to Hückel's scheme. Consider just two atomic orbitals, $|A\rangle$ and $|B\rangle$, associated with different atoms A and B. As long as the two atoms are far apart, an electron prepared in the state $|A\rangle$ (or $|B\rangle$), will stay in it indefinitely. The formation of a bond is essentially due to the 'tunneling' of the electron from A to B, when B approaches A so that the state $|A\rangle$ no longer has an infinite lifetime. At ordinary bond distances, the 'period' of the resulting electron oscillation is actually so short that, even if the electron could be prepared in $|A\rangle$, it would not stay there for a physically significant length of time. Thus the electron is delocalized over the two orbitals $|A\rangle$ and $|B\rangle$, and this is the reason why a stationary state of the AB system is a linear combination of $|A\rangle$ and $|B\rangle$.

The main difficulties with this picture arise when one considers, say, two electrons occupying the resulting molecular orbital. For one thing, either the orbital energy is the energy of one of the electrons, and then, in principle, twice the orbital energy is *not* the energy of the two-electron system; or the orbital energy is half the total energy, and then it does not correspond to a wave function even roughly resembling the 'correct' (Hartree-Fock) wave function. In practice, this apparent dilemma is less serious than it seems because the term R_0 representing the core repulsion may be shown to be of the same order of magnitude as the excess electron-electron interaction (Del Re and Parr 1963, Fischer-Hjalmars 1966); thus, in the assumption of complete equivalence of core repulsion and excess-electron repulsion, the sum of orbital energies does represent the correct energy of the system (which is thus given by Eq.(2.2.7) without R_0):

$$E_{Hü} = \sum n_j \varepsilon_j = \sum p_{\mu\mu}\alpha_\mu + \sum_{\substack{\mu,\nu \\ \text{nearest neighbours}}}' p_{\mu\nu}\beta_{\mu\nu} \quad . \tag{3.1.3}$$

This assumption has often been made with both the Hückel method and the so-called Extended Hückel method (Hoffmann 1963), and does not seem to be particularly bad. However, it must be remembered that it can only be valid at standard intermolecular distances. For this reason, and because it gives the wrong dissociation limits, the validity of the Hückel method is limited to standard geometries. We shall discuss its possible extension later.

As is well known, the greatest success of the original Hückel method lay in the explanation of conjugation, not in the treatment of isolated bonds, which belonged essentially to the quantum theory of diatomic molecules. The new idea was that the existence of non-vanishing transfer integrals between several atoms, and not just the two partners of a bond, would generate some sort of extended free-electron path (or, alternatively, a sort of many-centre bond), whereby an electron is delocalized over several atoms. This is why the Hückel method provides an effective first-order model for conjugated organic molecules as well as for metals.

A hierarchy of models - purely topological model, geometry-adapted Hückel model, Hückel model for molecules with heteroatoms - can be defined, and essentially amounts to simplified schemes for obtaining a qualitative quantum mechanical understanding of the consequences of (i) electron transferability from site to site, (ii) slight perturbations of the ideal nearest-neighbour coupling scheme, and (iii) the existence of local potential-energy wells - or barriers - corresponding to heteroatoms ("impurities"). A most important aspect of the Hückel model is that it treats the heteroatom-induced electron transfer from site to site and from bond to bond simply as the result of the orbital-coupling scheme. This effect - which the organic chemists call 'mesomeric' - is to be compared with the 'inductive' effect, consisting in bond polarization.

The most effective field of application of the Hückel method was historically the theory of conjugated hydrocarbons. The reasons for the remarkable success are of two types. In the first place, the questions to be answered with hydrocarbons were mainly of a semiquantitative nature: In particular, the important points were the stabilization of a system of conjugated double bonds depending on its topology; the changes in colour (more precisely, in the position of the strongest bond of UV and visible absorption spectra); the directing power of substituents with respect to addition reactions. In the second place, the conjugated hydrocarbons have substantially equivalent atomic orbitals, so that, if the zero of energy is chosen as the common α-value and the energy unit as the common value of $\beta_{\mu\nu}$ between directly linked atoms, the Hamiltonian (3.1.1) reduces to

$$\hat{H}^{(s)}_{red} = \sum_{\substack{\text{linked}\\ \text{atoms}}}{}' \hat{a}^{+}_{\mu s} \hat{a}_{\nu s} \qquad (3.1.4)$$

and is represented by the 'topological matrix' of the system. A number of well-known theorems (in particular those concerning alternant systems) apply to this expression.

An illustration of how the simple Hückel method can be used to gain some insight into the differences in stability of conjugated systems differing by their topologies is given in Table 3.2. The side values are 'stabilization' energies

Table 3.2. Interaction energies in some hydrocarbon π systems [1] (units $\beta_{cc} = 1$; interacting units separated by broken lines).

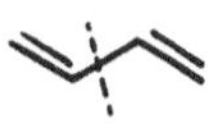	.175	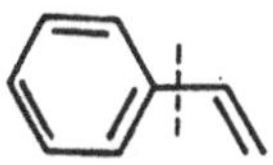	.143
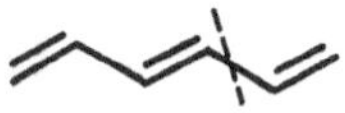	.212	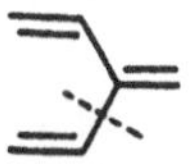	.147
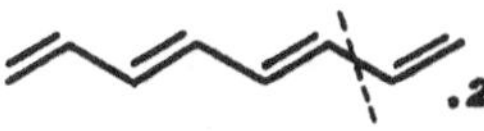	.262	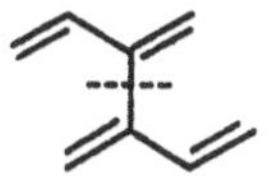	.124
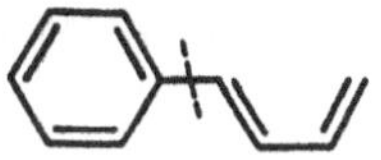	.208	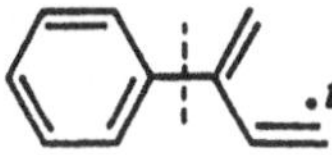	.120
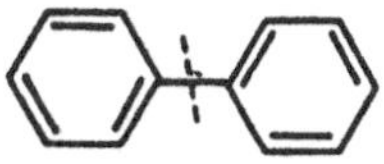	.118	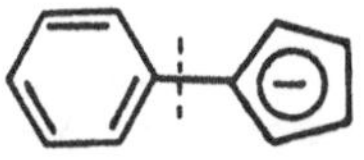	.121

[1]: The interaction energy is defined as the difference between the energy of the whole system and the sum of the energies of the separate units.

computed from the Hamiltonian (3.1.4). The most evident conclusion is that, according to the Hückel method, a system of conjugated π -bonds forming a linear chain is much more stable than a branched arrangement of the same bonds.

Conclusions of this kind are very interesting even if they are qualitative, provided, of course, that they are valid beyond the limits of the Hückel model. Care should be taken at least to estimate the chances that it is so.

In the case of Table 3.2 it can be shown that, if the basic assumptions (3.1.2) and (3.1.3) are granted, the nearest-neighbour approximation and the neglect of differences in bond distances and local environment of atoms only contribute second-order corrections to the energy.

Let us set

$$\delta_\mu = \frac{\alpha_\mu - \alpha}{\beta}, \quad \bar{\eta}_{\mu\nu} = \eta_{\mu\nu} - 1, \quad \eta_{\mu\nu} = \frac{\beta_{\mu\nu}}{\beta}, \tag{3.1.5}$$

where α and β are the diagonal and off-diagonal elements (of the 'ideal' Hamiltonian matrix) taken as the zero-point and as the energy unit, respectively, in constructing $\hat{H}^{(s)}_{red}$ of Eq. (3.1.4). Then the quantities thus defined are small, and, according to perturbation theory, the charge-bond-order matrix $\underset{\sim}{P}$ defined by (2.2.8) is very closely the same as that obtained from (3.1.4), *i.e.* by setting all δ_μ's and $\bar{\eta}_{\mu\nu}$'s equal to zero. The result for energy is thus

$$E_{Hü} = \left[n\alpha + \beta \sum_{\substack{\text{linked}\\ \text{atoms}}} p_{\mu\nu} \right] + \beta \left[\sum p_{\mu\mu}\delta_\mu + \sum_{\substack{\text{linked}\\ \text{atoms}}} p_{\mu\nu}\bar{\eta}_{\mu\nu} \right] \tag{3.1.6}$$

the first square bracket being the energy E_{top} obtained from Eq. (3.1.4) (n number of electrons).

The analysis of the results obtained for conjugated molecules containing heteroatoms

Table 3.3. Same as Table 3.2 with ● = nitrogen [1].

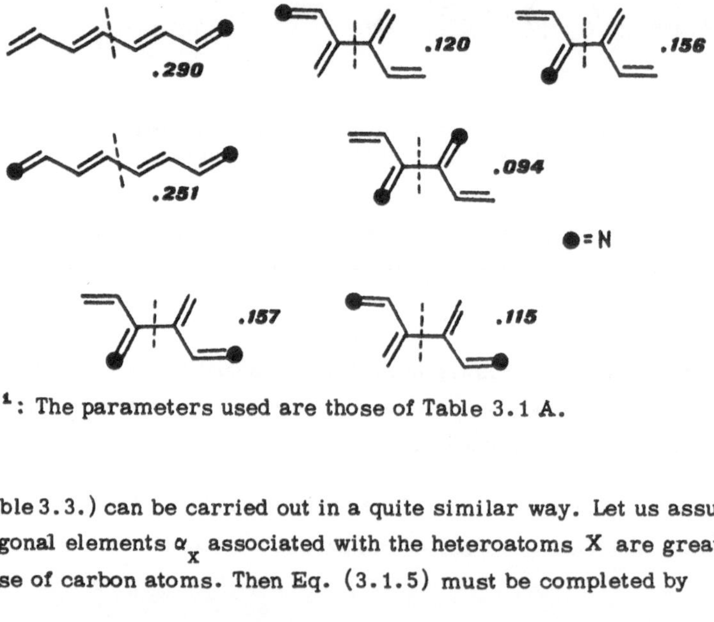

[1]: The parameters used are those of Table 3.1 A.

(Table 3.3.) can be carried out in a quite similar way. Let us assume that only the off-diagonal elements α_X associated with the heteroatoms X are greatly different from those of carbon atoms. Then Eq. (3.1.5) must be completed by

$$\alpha_x = \left(\delta_x^0 + \delta_x\right)\beta + \alpha, \tag{3.1.7}$$

and the energy of the system for a given occupation scheme is, to first order,

$$E_{Hü}^{(X)} = E_{top} + \beta \sum_{x} \delta_x^0 p_{xx}^0 + \beta \left[\sum_{x} \sum_{\substack{\mu \\ \text{linked} \\ \text{to } x}} \Delta p_{\mu x} + \sum_{x} \delta_x^0 \Delta p_{xx} + \sum_{\substack{\mu,\nu \\ \text{linked}}} \bar{\eta}_{\mu\nu} p_{\mu\nu}^0 \right] \tag{3.1.8}$$

where p_{xx}^0 and $p_{\mu\nu}^0$ denote values obtained for the 'parent' hydrocarbon with the Hamiltonian (3.1.4). Perturbation theory makes the estimation of $\Delta p_{\mu x}$ and Δp_{xx} possible. We shall not discuss the resulting expressions, but we emphasize that they are the way to a detailed analysis of the role of different effects on the π-electron energy of a heteroatomic conjugated system: Starting with a model where only topology is considered, changes in geometry, in environment, and in nature of the atoms are successively introduced. In the same way also the nearest-neighbour approximation can be removed by a correction which represents the 'far-neighbour effect'.

The same kind of analysis can be carried out on other properties of conjugated molecules containing heteroatoms.

As should be clear from the above considerations, the Hückel method and its elaborations provide a powerful tool for the resolution of molecular properties into contributions resulting from distinct features of the structure of molecules and crystals, in particular effects like long-range electron transfer (mesomeric effect). Of course, the basic condition that a one-electron scheme be associated with a basis containing one orbital per atom is essential in this first type of scheme. With that condition, even an ab-initio SCF scheme could (and perhaps should) be discussed in terms of the simple hierarchy of effects suggested by the Hückel method.

Several reviews of the Hückel method have appeared, including well-known textbooks, like Pullman and Pullman's (1952), Streitwieser's (1961), Daudel, Lefebvre and Moser's (1959). R. D. Brown's review (1952) was written at the time of the highest popularity of the Hückel method. Perhaps the most exhaustive study is the excellent textbook of Heilbronner and Bock (1970), whose third volume is entirely devoted to practical examples, with special emphasis on hydrocarbons (i.e. on the topological model). As is illustrated by Table 3.1 the Hückel method is still used by chemists and physical chemists working with aromatic molecules, so that recent applications can be found here and there in the chemical literature. Another wide field of application is found in solid state physics, as will be mentioned later.

As regards the present status of the Hückel method, and the question whether it can still render useful services, a complete picture can only be obtained by taking into account the considerations made later in the present chapter, because certain points are discussed in connection with σ systems. We should like to suggest here that the Hückel model is the appropriate one for any problem that can be stated in the form: Can experimental trends (in the static or dynamic properties) be given a reasonable explanation in terms of the electron mobility associated with conjugated bonds, as is done for the aromaticity of benzene?

The important point here is that we are not looking for rigorous quantum mechanical interpretations nor for accurate quantitative predictions; we are just asking *whether certain trends can* be attributed to conjugation. Therefore, use of a more complicated model would mean obscuring the required information by mixing it with additional information. Indeed, even the choice of parameters may be relatively unimportant, as was pointed out to one of the present authors by H. C. Longuet-Higgins in 1956, in connection with the problem of explaining why, whereas pyridine-N-oxide II has an observed dipole moment about 2D larger than pyridine, benzonitrile-N-oxide IV had been found to have practically the same dipole moment as benzonitrile III. Why should

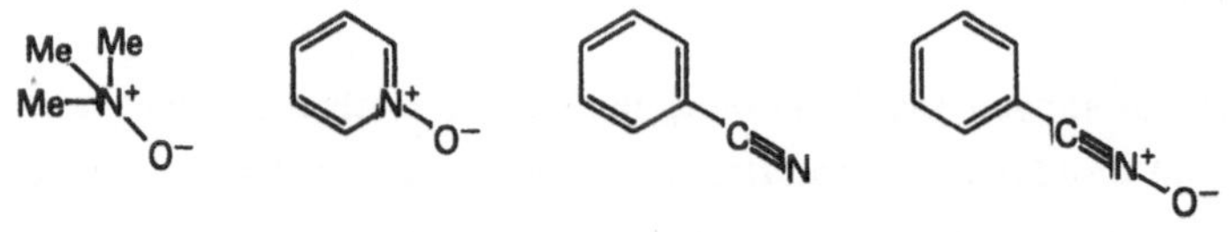

I II III IV

a dative N - 0 bond show practically a zero bond moment in IV, when its moment was over 4D in trimethylamine-N-oxyde I? Could such a striking effect be due to back donation by 0 to the π-system of III?

It seemed that a Hückel calculation could provide an answer, and the parameters used were the simplest ones, namely $\delta = 1$ for nitrogen, $\delta = 2$ for oxygen, $\eta = 1$ for all bonds. Such a choice is a very poor one for quantitative purposes, but it was sufficient, for it predicted that the increase in dipole moment from pyridine to II would be far larger than the increase from III to IV. Thus, although the computed dipole moments were off by a few percent in pyridine and by 100 % in IV, the answer was unambiguous: the apparent zero moment of the $N^+ - O^-$ bond in benzonitriloxide could be accounted for by a mesomeric effect (Del Re 1957).

Further illustrations of the same kind of applications are relatively frequent especially in the literature of the sixties, although they are outnumbered by studies just aimed to prove that the Hückel method can be made to reproduce accurately the most diverse properties of molecules. As regards circumstances under which improvements of the Hückel model are really needed, an illustration will be given in chapter 5 using the example of azulene.

3.2. A 'Naive' Method for σ Electrons

The Hückel method was essentially intended to explain the peculiar properties of conjugated systems of double bonds in organic molecules, and therefore for a long time no similar method was proposed for σ bonds. A few attempts in that direction were made as late as 1955 by Sandorfy and by Fukui in 1961. In 1958 one of us proposed a method (Del Re 1958, 1964) which, because of its simplicity and ease of computation, retained some popularity even when the so called 'all valence-electrons' methods entered the arena. We shall discuss that method because it is to a large extent the counterpart of the Hückel method: whereas the latter describes the so-called mesomeric effects, the former describes the so-called inductive effects of saturated molecules (indeed, it was proposed as a procedure to estimate *in situ* dipole moments).

3.2.1. The inductive effect and atomic parameters. - The inductive effect was introduced in connection with reaction mechanisms (Ingold, 1953) to explain the influence of substituents on the reactivities of certain sites of saturated molecules. A characteristic of that effect is that, in contrast to the mesomeric effect, it decays exponentially with the distance from the perturbing site.

On the basis of that consideration, and extending a practice already adopted for π systems (cf. 3.1.1), the criterion was adopted that inductive effects would be described by adding a suitable fraction of the atomic parameter δ_B of Eq. (3.1.5) of a site B adjacent to a given site A to the atomic parameter δ_A of the latter. In order to introduce a certain amount of long range effects it was further assumed that the above rule should hold for *any* site: for instance, in a chain of three atoms A - B - C, δ_B would contain a fraction $\gamma_{B(A)}\delta_A$ of δ_A, and δ_C would contain a fraction $\gamma_{C(B)}\delta_B$ of δ_B. As γ must be much smaller than unity, the fraction of δ_A 'transmitted' to C is thus a second-order correction, but may be quite significant, especially in the practical cases where an atom A is influenced by several neighbours.

The coefficient $\gamma_{B(A)} \equiv \gamma_{BA}$ will be called here the 'transfer coefficient'. In fact, the scheme just described amounts to treating each δ as a signal transferred from one atom to its neighbours, and back. With the symbols used above, the expression for an atomic parameter will be

$$\delta_A = \delta_A^0 + \sum_{B \neq A} \gamma_{A(B)} \delta_B \qquad (3.2.1)$$

δ_A^0 being the atomic parameter corresponding to the ideal situation when A is not influenced by its neighbours.

As will be shown later, Eq. (3.2.1) is but a simplified form of an SCF procedure. In fact, the determination of the atomic parameter of an atom A demands solving the system of equations obtained by making the subscript A of Eq. (3.2.1) run through all the atoms of the given molecule *(vide infra)*.

In the case of a diatomic molecule, solving the system (3.2.1) corresponds to finding the limiting values of a process whereby atom A modifies δ_B, the modified B modifies A, and so on to convergency. For example, if the transfer coefficients are .04 in the direction A to B, and .03 in the opposite direction, we have the sequence (up to 10^{-7}):

A	δ_A^0,	$\delta_A^0 + 0.03\ \delta_B^0 + 0.0012\ \delta_A^0$,
B	$\delta_B^0 + 0.04\ \delta_A^0$,	$\delta_B^0 + 0.04\ \delta_A^0 + 0.0012\ \delta_B^0 + 0.000048\ \delta_A^0$,
A		$\delta_A^0 + 0.03\ \delta_B^0 + 0.0012\ \delta_A^0 + 0.000036\ \delta_B^0$,
B		$\delta_B^0 + 0.04\ \delta_A^0 + 0.0012\ \delta_B^0 + 0.000048\ \delta_A^0$,

whereby the difference in the atomic parameters, which is the important quantity for the polarity of a bond, becomes

$$.971164\left(\delta_A^0 - \delta_B^0\right) - .010012\ \delta_A^0 .$$

3.2.2. <u>A one-electron Hamiltonian for localized bonds.</u> - The model Hamiltonian to be associated with the method under study is referred to a basis of more than one orbital per atom: Every atom is assumed to contribute to the AO basis a number of orbitals equal to its valency. Nevertheless, the atom-bond picture is preserved by assuming that the atomic parameter associated with any given atom is independent of the particular bond *(viz.,* in this case, of the particular atomic orbital) under consideration, whereas transfer coefficients and bond parameters are associated with pairs of atomic orbitals making up a bond.

For a given atom A we write the orbital forming the bond with the neighbour B_A as $|A(B)\rangle$. Then the model Hamiltonian can be written

$$\hat{H}_{DR} = H_0 + \hat{H}_{DR}^{(\uparrow)} + \hat{H}_{DR}^{(\downarrow)} \tag{3.2.2}$$

$$\hat{H}_{DR}^{(s)} = \sum_{A}\sum_{B_A}\left[\delta_A \hat{n}_{A(B),s} + \eta_{AB}\hat{a}^{+}_{A(B)s}\hat{a}_{B(A)s}\right] \tag{3.2.3}$$

This can also be decomposed into a sum of bond Hamiltonians:

$$\hat{H}_{DR}^{(s)} = \sum_{\substack{\text{bonds}\\(AB)}}\left[\delta_A \hat{n}_{A(B)s} + \delta_B \hat{n}_{B(A)s} + \eta_{AB}\left(\hat{a}^{+}_{A(B)s}\hat{a}_{B(A)s} + \hat{a}^{+}_{B(A)s}\hat{a}_{A(B)s}\right)\right] \tag{3.2.4}$$

Of course, δ_A and δ_B must be computed according to Eq. (3.2.1), and that is the only way in which the presence of other bonds affects any given bond AB.

The Hamiltonian (3.2.4) is valid both in an orthogonal and in a non-orthogonal basis (let aside the fact that in the latter case $\hat{n}$ is no longer the number operator). The original method assumed orthogonal orbitals on the nuclei. It must be shown that such an assumption does not imply an internal contradiction (Del Re 1960, 1976); it is found that some care in the expression for orbital energies may be necessary, but there is no problem with charges.

3.2.3. Atomic parameters and the electronegativities of atoms in situ. - The parameters to be used in the Del Re method have been the subject of many studies. Probably the most extensive one has been carried out by Vandorffy (1972) who has taken into account hybridization, so as to provide a complete set of parameters valid also for the σ systems of unsaturated molecules (a problem treated already by Berthod and A. Pullman in 1965, 1967). A list of parameters is given in Table 3.4.

Vandorffy found that the δ^0 values could be correlated quite well with electronegativities by means of the equation

$$\delta_A^0 = k\,\frac{\chi_A - \chi^0}{\chi_A + \chi^0} \tag{3.2.5}$$

In the original study of Vandorffy the χ's were Pauling electronegativities and χ^0 and k were adjusted on the parameter values of hydrogen and sp^3 carbon (column 4 Table 3.4). We have recalculated the δ^0's using the Mulliken electronegativities of Table 1.3 and using as reference values those of tetrahedral carbon and of tetrahedral fluorine, and have obtained the values

$$k = 1.2797, \quad \chi^0 = 6.9461 \tag{3.2.6}$$

which give the results shown in column 5 of Table 3.4. Equation (3.2.5) provides δ^0 values for all hybridization, and even smoothes out certain anomalies; in our opinion it should be preferred to others even if it did not give as good agreement with experimental dipole moments, because it reduces drastically the number of adjustable parameters. What is perhaps more important, it reduces the problem of the physical significance of the δ^0 parameters to that of the physical significance of electronegativities; for (3.2.5), in agreement with a remark by Majee and Gupta (1972), essentially states that the δ^0's are another way of expressing valence-state electronegativities (cf. also Berthod and A. Pullman 1965). The case of sulphur is an interesting example; the value 0.07 estimated from dipole moments by Del Re, B. Pullman, and Yonezawa (1963) corresponds to the electronegativity of a very slightly hybridized valence state. A formal analysis is given in Sec. 4.4.2.

An analysis of the other parameters (Table 3.) is also possible, and would lead to interesting conclusions. We note only that, given the γ values, the η values may be adjusted to the electronegativity scale of column 5 of Table 3 so as to roughly obey the condition that

$$C_{AB} = \frac{\delta_A^0 - \delta_B^0}{2(1 - \bar{\gamma}_{AB})\eta_{AB}} \tag{3.2.7}$$

should be same for different choices of the δ^0's, where $\bar{\gamma}_{AB}$ is the average of γ_{AB} and γ_{BA}.

It may be argued that, in spite of the correlation with electronegativities, the parameters in question are too many and that this makes the model practically tautological. The answer is that the number of possible combinations of the various types of bonds is enormous, so that even if each type of bond or of atom is assigned a relatively large number of parameters, the molecules to which those parameters have to be applied are by several orders of magnitude more numerous than the parameters.

Nevertheless, the parameter question is particularly evident here, and it is convenient to comment on it.

Table 3.4. δ^{o}-values and Correlation with Electronegativities.

ATOM	(1)	(2)[a]	(3)	(4)	(5)[a]	(6)	Ref. to (4)
H		2.21	7.17	0.00	0.00	0.020	(1) (2)
C	(tetetete)	2.48	7.75	0.07	0.07	0.070	(1) (2)
C	(trtrtr π)	2.75	8.79	0.12	0.13	0.150	(2)
N	(te^2tetete)	3.68	11.54	0.24	0.30	0.318	(1) (2)
	(trtrtr π^2)	3.94	12.31	0.30	0.34	0.356	(2)
	(tr^2trtr π)	4.13	12.87	0.38	0.37	0.383	(2)
0	(te^2te^2tete)	4.93	15.25	0.40	0.46	0.479	(1)
	(tr^2trtr π^2)	5.43	16.73	0.60	0.51	0.529	(2)
F	($te^2te^2te^2$te)	5.94	18.16[b]	0.57	0.56	0.572	(1) (2)
Si	(tetetete)	----	7.29	----	----	0.031	
Si	(sppp)	1.82	6.00	- 0.10	- 0.12	0.094	(3)
P	(te^2tetete)	----	8.90	----	----	0.158	
	(trtrtr π)	----	9.47	----	----	0.197	
	(tr^2trtr π)	----	9.66	----	----	0.209	
S	(tr^2trtr π^2)	----	10.88	----	----	0.282	
	(te^2te^2tete)	----	10.13	0.07	----	0.239	(4)
Cl	(s^2p^2pp)	----	7.38	----	----	0.040	
	($te^2te^2te^2$te)	4.01	12.46[b]	0.35	0.35	0.364	(1) (2)

[a] - from Vandorffy, 1972.
[b] - Electronegativity from Table 1.3 according to Pongor (cited in Vandorffy, 1972).

EXPLANATION: column (1): hybridization state
column (2): Pauling electronegativity.
column (3): Mulliken electronegativity (cf. Table 1.3).
column (4): current δ^{o}-values
column (5): δ^{o}-values from Pauling electronegativities.
column (6): δ^{o}-values from Mulliken electronegativities.

REFERENCES: (1) Del Re, 1958.
(2) Berthod *et al.*, 1967
(3) Nagy *et al.*, 1970
(4) Del Re, 1964.

As has been remarked in connection with the Hückel method, quantitative improvements do impair the quality of the model. The Hückel scheme for hydrocarbons is based exclusively on the topological arrangement of bonds, whereas extension to heterocycles introduces additional effects (through *ad hoc* parameters) which mix in to make the picture more complicated. Here we have a situation where the basic effect is not a topological one but the juxtaposition of bonds having different polarizabilities. This is why in the σ case quantitative aspects play a more important role than in π systems. Nevertheless, here too we have an essentially qualitative aspect, namely the way in which the presence of a given bond is felt by the rest of the molecule. In other words, although the parametrization is rather heavy, it does not really concern the description of inductive effects; it is just needed to represent the building blocks of the molecule, much as in certain semiempirical and *ab initio* methods the atoms are represented by *ad hoc* sets of orbital exponents and quantum numbers.

The above consideration, of course, does not answer the question whether or not a bond should be represented in a simpler way, *viz.* by as few parameters as possible. Both a negative and a positive answer can be defended. The effectiveness of the model in detecting trends or in defining effects is enhanced, of course, when the number of parameters is reduced: but the result may be completely artificial if it does not isolate a major effect. Therefore, as a general rule, the various parameters should be studied individually before deciding which of them, if any, should be chosen to account for some major effect.

3.2.4. <u>Numerical details.</u> - To describe the method under study at a more concrete level, we reformulate the above presentation as follows. Consider again Eqs. (3.2.3) and (3.2.4). The diagonal and off-diagonal elements of the matrix **H** representing the one-electron Hamiltonian are the coefficients of $\hat{n}^+_{\mu s}$ and $\hat{a}^+_{\mu s}\hat{a}_{\nu s}(\mu \neq \nu)$, respectively. Now, the inductive effect may be interpreted as a change in polarization of a given bond brought about by the presence of neighbouring bonds. The fact that no charge transfer is included in the inductive effect is reflected in the structure of the matrix **H** (which is block-factorized into "bond Hamiltonians" $\mathbf{H}_{bond}$): the inductive effects are just introduced through some change in the matrix elements of the bond Hamiltonian, and not - as is the case with conjugation - through non vanishing off-diagonal elements connecting blocks associated with different bonds.

The maximum of simplicity is obtained, first of all, by assuming that the hybrid orbitals entering a bond orbital are orthogonal to each other. This is somewhat strange at first sight, but is actually completely appropriate to the 2×2 problem, because then the off-diagonal part of the Hamiltonian matrix is always proportional to overlap, and a simple transformation makes the orthogonalized and the non-orthogonal problems equivalent.

Next, we assume that the diagonal elements of the bond Hamiltonian $\mathbf{H}_{bond}$ roughly measure the electronegativities of the atoms involved - an assumption already introduced in the standard Hückel method - whereas the off-diagonal elements measure at the same time the strength and the resistance to polarization of the bond. Calling X and Y the two atoms, and keeping in mind that in a saturated molecule we have one bond per atom pair, we write:

$$\mathbf{H}_{bond} = \begin{vmatrix} H_{xx} & H_{xy} \\ H_{xy} & H_{yy} \end{vmatrix} = \alpha \begin{vmatrix} 1 & 0 \\ 0 & 1 \end{vmatrix} + \beta \begin{vmatrix} \delta_x & \eta_{xy} \\ \eta_{xy} & \delta_y \end{vmatrix} \tag{3.2.8}$$

which defines the dimensionless parameters δ_x, δ_y, and η_{xy} in units β and with an energy zero-point exactly as is customary in the Hückel method. As is well known, with that unit and that zero-point, $\mathbf{H}_{bond}$ can be replaced for all practical purposes by the matrix

$$\mathbf{H}_{bond} = \begin{vmatrix} \delta_x & \eta_{xy} \\ \eta_{xy} & \delta_y \end{vmatrix} \tag{3.2.9}$$

whose eigenvectors give the coefficients of the molecular orbital. With the necessary reservations, we can identify the atomic charges with the defect or excess electron populations

$$q_{x(y)} = 1 - 2\,c^2_{x(y)}, \quad q_{x(y)} = -\,q_{y(x)} \tag{3.2.10}$$

where the partner of X is specified because atom X may form several bonds. In fact, its total charge will be

$$q_x = \sum_y q_{x(y)} \tag{3.2.11}$$

where Y denotes all the atoms directly linked to X. For instance, in methane, for a CH bond one finds:

$$\mathbf{H}_{CH} = \begin{vmatrix} 0.1343 & 1 \\ 1 & 0.0538 \end{vmatrix} \tag{3.2.12}$$

if the parameters are evaluated according to Table 3.2.

Consider now Eq. 3.2.1.

Since the δ's appearing on the right hand side of that equation are determined by equations of the same type, the set of the δ's is the solution of a linear system of equation which, in matrix formalism, can be written

$$\Gamma\Delta = \Delta^0 \qquad (3.2.13)$$

Γ being a non-symmetric matrix with diagonal elements 1 and off-diagonal elements 0 for non-bonded atom pairs. The numerical form of Eq. (3.2.13) for methanol, with atoms taken in the order H(of CH_3), C, O, H, is

$$\begin{array}{cccccc} H & H & H & C & O & H \\ \end{array}$$
$$\left|\begin{array}{cccccc} 1 & 0 & 0 & -.4 & 0 & 0 \\ 0 & 1 & 0 & -.4 & 0 & 0 \\ 0 & 0 & 1 & -.4 & 0 & 0 \\ -.3 & -.3 & -.3 & 1 & -.1 & 0 \\ 0 & 0 & 0 & -.1 & 1 & -.3 \\ 0 & 0 & 0 & 0 & -.4 & 1 \end{array}\right| \left|\begin{array}{c} \delta_{H_1} \\ \delta_{H_2} \\ \delta_{H_3} \\ \delta_C \\ \delta_O \\ \delta_H \end{array}\right| = \left|\begin{array}{c} 0 \\ 0 \\ 0 \\ 0.07 \\ 0.40 \\ 0 \end{array}\right| \qquad (3.2.14)$$

The parameters used here are from the set originally chosen in order to improve the bond moment description of saturated molecules by allowing for inductive effects. In the present context, only the fact that inductive effects appear in the net charges derived according to the above scheme is important; it is illustrated by the examples of methyl-substituted water and amines (Table 3.5).

Table 3.5. δ-values and Charges for the Molecule H_2FCOH (δ^0-values from Table 3.4).

ATOM	δ^a	q^b	δ^c	q^d
H (of CH_2)	0.0934	0.0699	0.1252	0.0688
F	0.5934	- 0.0968	0.5983	- 0.0902
C	0.2335	0.0860	0.2631	0.0861
O	0.4811	- 0.4347	0.5810	- 0.4246
H (of OH)	0.1924	0.3054	0.2524	0.2911

[a] - from Table 3.4, column 4.
[b] - charges from column (a) with $\eta_{CH} = 1$; $\eta_{CO} = 0.95$; $\eta_{CF} = 1.85$; $\eta_{OH} = 0.45$.
[c] - from Table 3.4, column 6.
[d] - charges from column (c) with $\eta_{CH} = 1$; $\eta_{CO} = 1.18$; $\eta_{CF} = 1.85$; $\eta_{OH} = 0.54$.

3.2.5. <u>Critical comments</u>. - The above method was developed at a time when only simple schemes were practicable. The questions that can be raised about its possible usefulness in the present state of quantum chemistry are of the same type as those raised in connection with the Hückel method. We formulate here four of those questions:

a) what about molecules with σ and π electrons?
b) have the results obtained any validity as compared with all-valence-electron semi-empirical and *ab initio* methods?
c) are the implicit assumptions of the method non-contradictory?
d) what kind of problems can be solved by the method in question despite its simplicity?

The case of molecules with localized and delocalized electrons may be reduced to that of a σ system where hybridization is different from that of a fully saturated molecule, plus a π system with suitably modified parameters (to take the polarization of the σ frame into account). Approaches based on that sort of considerations were proposed among others by Berthod *et al.* (1967) and by Momicchioli and Del Re (1969). The validity and the usefulness of the σ - π separation is, of course, prejudicial to any decision concerning the significance of those approaches. As has been mentioned, that problem is quite delicate, and it is difficult to resist the temptation of suppressing it by having recourse to an all-valence-electron approach. However, separations are useful to find out special features that may be responsible for given properties. For instance, it is both reasonable and useful to interpret optical spectra in terms of π and n electrons only; let us remember that all the members of the immense class of dyes owe their colour to systems of conjugated bonds. Therefore, treatments separating σ and π electrons should be taken very seriously, especially in connection with spectroscopy. The well-known objection that it is difficult to define π systems in non-planar conjugated systems (as in twisted biphenyl) is probably the most serious one; but circumvention of the corresponding physical problem by just giving the σ - π separation up is not a fair answer.

The σ method just described was originally designed to yield net atomic charges. It was used later for conformational studies, to predict reactivity trends (Gupta and Majee, 1972), and to define theoretical bond energies (Barone *et al.* 1979). The published investigations suggest that, possibly with minor parameter adjustments, the predicted trends are quite reasonable. The results are satisfactory at least to first order, and it can be said that the corresponding properties of saturated molecules and σ-frames can be explained in terms of inductive effects, interpreted according to the above method as electronegativity adjustments. As in the case of the Hückel method and mesomeric effects, such a claim does not mean that strictly quantitative results are 100 % reliable; even if the necessary parameters are considered final (which is not the case, especially as regards the η values), the method can at best provide results which would be quantitatively correct if no other effect were there.

Thus, comparisons with methods that are hopefully more complete may be understood as attempts to test the importance of an inductive-effect description of molecules. Such is the case, for instance, with a careful investigation by Fliszar (1972), who concluded that the net charges found by various methods, including *ab initio* ones, show a satisfactory parallelism.

3.3.3. Core repulsions. Applications. - The second difficulty mentioned in the introduction to this section concerns the so-called core repulsions. We shall not discuss them in general here, because they will be treated in sec. 4.2. However, we mention that they can perhaps be neglected as long as the distances between nearest neighbours are assigned standard values, because then suitable choices of the hopping integrals are sufficient to take those forces into account: but it is no longer so, because of differences in the laws of variation, when different distances are to be considered (Del Re *et al.* 1977). However, in spite of different ways of handling this difficulty, interesting results have been obtained especially in the treatment of small particles by the Hückel method. Julg *et al.* (1974) and Cyrot-Lackmann *et al.* (1977) have studied the stabilities of different shapes and symmetries of clusters with increasing sizes. A very important result of those studies is that the arrangements that are most stable when the atoms are few are not the most stable ones when the number of atoms is greater than a critical value. Julg *et al.* (1977) have studied by the Hückel method the distortions induced in otherwise regular clusters (in particular planar ones) by the boundaries. The particular version of the Hückel method used there could be classified as a semiempirical SCF method, because it is iterative: hopping integrals β are determined from the distances, the distances are determined from the bond orders, the bond orders are obtained by diagonalizing a matrix which contains in general another set of β's; the procedure converges when the β's used to determine the bond orders and the β's obtained from the bond orders via the distances are the same. We consider that procedure still a Hückel method because the computation *for a given geometry* is not iterative; but, of course, a different point of view would have no practical consequences. A more interesting question is the general validity of the results of those calculations. We emphasize again that model studies are not expected to predict theoretically unquestionable facts; they are expected to provide schemes of interpretation and to suggest trends corresponding to the particular effects that are included in the model, e.g. the nearest-neighbour coupling and the one-electron scheme. A very serious methodological question arising in this connection is: Is it worth-while to consider more sophisticated semi-empirical methods, and what can we expect to learn from them? We shall return to this question after having introduced the Extended Hückel Method (EHT).

3.3. The Hückel Method for All Valence Electrons: the Tight-binding (TB) Approach of Solid-state Physics.

Whereas the π systems of conjugated molecules are the traditional field of application of the Hückel method in chemistry, the same method has seen a number of interesting applications to all the valence electrons in the case of crystals, especially metals, the reason being, of course, that in those systems a delocalization of all the valence electrons fully analogous to that of π electrons must be expected.

The assumptions made are the same as in the standard Hückel method: one orbital per atom, orthogonality of the basis, nearest-neighbour approximation, standard value of the transfer (or 'bond', or 'hopping') integral β for a given atom pair and distance (Ducastelle and Cyrot-Lackmann, 1970). However, there are some additional problems. First, what is a nearest neighbour in the case of a crystal? Second, what repulsive forces must be introduced to counteract the attractive nature of the electron-nucleus forces? Third, what is the situation when the atoms under study, instead of being alkali-like (*i.e.* atoms sharing only one orbital), are, say, transition metal atoms, entering the crystal orbitals with their *s*, *p*, *d* orbitals?

3.3.1. The nearest-neighbour approximation. - The second and third difficulties are not especially important when model studies not concerning relaxation effects are carried out, but the first one is generally serious. To see it, consider a square array of 9 atoms simulating a plane of a cubic crystal. At first sight, one would consider as first neighbours just the atoms connected by the sides of a square: but, considering that the diagonals are only 41 % longer than the sides, and that the interactions along them are necessarily important (otherwise the system would not be rigid), one can doubt that the simplest Hückel recipe - equal β's for nearest neighbour bonds, zero elsewhere - is valid *sic et simpliciter*. On the other hand, calculations made with suitable β's on the diagonals give results comparable to those of a strict nearest-neighbour approximation (n.n.a). The situation can be analyzed in terms of equation (2.2.7), using - on the basis of perturbation theory - bond orders derived for the strict n.n.a. The change in energy due to having taken the diagonal into account is then

$$\Delta E_{diag} \simeq \sum_{J,K} P_{diag}^{JK} \beta_{diag} \qquad (3.3.1)$$

with p^{JK}_{diag} computed from the simpler case, for atoms J and K connected by a diagonal. In the case of a single square that bond order is zero; in general, the quality of the approximation neglecting the diagonal may be judged from (3.3.1).

3.3.2. The case of many orbitals per atom. - In certain cases - like, say, the *d*-band of transition metals - it is not possible to assume that there is a single orbital per atom; then further specification of the various off-diagonal elements of the Hamiltonian matrix is necessary. In the case of the *d* orbitals in an f.c.c. crystal Ducastelle and Cyrot-Lackmann (1970) have suggested that the intra-atomic off-diagonal elements are negligible, so that, in the n.n.a., the computation of band shapes breaks down into the three bands $d\sigma$, $d\pi$, and $d\delta$ which are controlled by one transfer integral each and are separately obtained by a Hückel calculation *sensu stricto*. Due to the crystal symmetry, the assumption that mixed transfer integrals are zero is quite acceptable as long as the nearest-neighbour approximation is accepted, but the considerations of the preceding subsection apply. As concerns intra-atomic transfer integrals, the crystal field can couple the *d* orbitals of different symmetries to one another. If that coupling is not neglected, it is not possible to speak of separated *d*-bands of the σ, π. and δ type.

Another example of the role of those intra-atomic elements is provided by diamond, silicon, and germanium. First of all, it would be possible to have separated *s* and *p* bands only if just two kinds of transfer integrals were different from zero in such solids, those between s orbitals of neighbouring atoms, and those between $p\sigma$ orbitals of neighbouring atoms. Chemically speaking, such a limitation is unjustified: in fact, the current model of diamond uses the standard sp^3 hybrids. Thus, in the n.n.a. there is but one hopping parameter between nearest neighbours; if the other parameters (between tetrahedral hybrids pointing in different directions) are neglected, localization is complete because the Hamiltonian matrix built according to the Hückel recipe appears to break down into two-by-two blocks. The neglected off-diagonal elements connecting those blocks include interatomic and intra-atomic terms; it is easy to convince oneself that the latter are more important. In fact, bands are obtained even in diamond by coupling different hybrids of the same atom (Leman and Friedel 1962).

A general formulation of the Hückel method for the most general crystal - *i.e.* a molecular crystal whose unit cell contains several different atoms each carrying several orbitals - was provided by Ladik (1965), who reduced the problem of determining the bands to diagonalization of a family of complex matrices depending on the wave-vector $\underset{\sim}{k}$ and of order $2n$, n being the number of AO's in the unit cell. Ladik also treated the case of pseudo-periodic systems.

3.3.3. The Hückel method as a pseudopotential scheme. - In sec. 2.3 we have briefly presented the scheme proposed by Durand for determining effective Hamiltonians, in particular pseudopotentials. A question may be asked: Can the Hückel method be interpreted in terms of a pseudopotential scheme with transferable atomic potentials? In other words, can we write for a molecule a reasonable one-electron Hamiltonian of the type

$$\hat{H}_{\text{eff}} = \hat{T} + \sum \hat{W}_{\text{ps,M}} \tag{3.3.2}$$

where $\hat{T}$ is the kinetic-energy operator, $\hat{W}_{\text{ps,M}}$ is a potential which depends only on the nature of the atom M and on its state of coordination. Nicolas and Durand (1979) have used the technique described in sec. 2.3 with the SCF molecular electronic Hamiltonian taken as the exact Hamiltonian and with methane as the reference molecule. They postulated for W the form of a non-local potential over Gaussian atomic orbitals. They obtained for carbon and hydrogen the following pseudopotentials:

$$\hat{W}_{\text{ps,C}} = -1.207 \left\{ |e^{-0.6r^2}\rangle\langle e^{-0.6r^2}| + 1.645 \left[|xe^{-1.12r^2}\rangle\langle xe^{-1.12r^2}| + \text{+ corresp. terms in y, z} \right] \right\}, \tag{3.3.3}$$

$$\hat{W}_{\text{ps,H}} = -1.300 |e^{-1.4r^2}\rangle\langle e^{-1.4r^2}|. \tag{3.3.4}$$

With these expressions the Hamiltonian (3.3.2) gives orbital energies very close to the SCF ones (obtained from a minimal STO-3G basis): In the first five saturated hydrocarbons the error is at most 0.017 a.u., but the mean square deviation is 0.01 in the worst case (isobutane); the total valence energies are correct to within 0.03 a.u.

We call attention to the peculiar mathematical form of non-local potentials: They are defined through their matrix elements. For instance, the matrix element of $\hat{W}_{\text{ps,H}}$ over two Gaussian s-type orbitals with orbital exponents ζ_1 and ζ_2, and centred on C and H_2, respectively, is

$$-1.300 \langle e^{-\zeta_1 r_c^2} | e^{-1.4 r_1^2} \rangle \langle e^{-1.4 r_1^2} | e^{-\zeta_2 r_h^2} \rangle,$$

and thus is defined in terms of overlap integrals.

A justification of the the Hückel method in terms of non-local potentials was attempted a long time ago by Simpson (1958), but to our knowledge it was not followed by practical applications. A justification of the extended Hückel method by direct comparison of orbital energies with SCF results was given by Newton *et al.* (1966). Durand's work appears to have realized a synthesis of the two approaches through the pseudopotential concept. It is also important that it is free from approximations like the n.n.a. (sec. 3.3.1) and the Wolfsberg-Helmholz formula (3.4.1).

3.4. All-Valence-Electron Treatments: the Extended Hückel Theory (EHT)

The emphasis on the fact that the Hückel model was so successful because is treated delocalized electrons hindered for a long time advances towards extension of the same scheme to both π and σ electrons. Indeed, the few authors who attempted anything of that sort (Sandorfy 1955; Fukui *et al.* 1962) confined themselves to hydrocarbons, thinking that otherwise the number of parameters would be far too large. It took a lot of 'semiempirical courage' (to use an expression coined by Zahradnik in 1968) to produce a Hückel method treating on the same footing all the valence orbitals of a molecule (Hoffmann 1963). The parameter problem was avoided by introducing for the bond integrals a formula connecting them to the overlap integrals $S_{\mu\nu} = \langle\mu|\nu\rangle$ and to the atomic parameters α_μ (essentially electronegativies), that is, the famous WH formula (Wolfsberg-Helmholz 1952)

$$\beta_{\mu\nu} = k\, S_{\mu\nu}\left(\frac{\alpha_\mu + \alpha_\nu}{2}\right). \qquad (k = 1.5\ldots 3) \qquad (3.4.1)$$

This formula has many attractive features: First of all, it depends only on the atomic orbitals under consideration for a given distance of their centres (because overlap is a function of the orbital exponents); second, it corresponds to the so-called Mulliken (1949) approximation for the matrix elements of an operator $\hat{Q}$ between two orbitals $|\mu\rangle$ and $|\nu\rangle$:

$$\langle\mu|\hat{Q}|\nu\rangle \simeq \frac{1}{2} S_{\mu\nu}\left(\langle\mu|\hat{Q}|\mu\rangle + \langle\nu|\hat{Q}|\nu\rangle\right) \qquad (3.4.2)$$

from which the WH approximation differs by a numerical factor; third, it implies a dependence on the internuclear distance that is found to be quite reasonable in hydrocarbons; finally, it does not involve the nearest-neighbour approximation (which would be difficult to specify in σ systems), though leaving the number of parameters the same or lower than a corresponding Hückel n.n.a. scheme.

On the other hand, as is mentioned in sec. 5.3.2, equation (3.4.1) rests on the choice of a specific form for the basis orbitals (Berthier *et al.* 1966). Moreover, (3.4.2) holds much better for potential energies than for kinetic energies (cf. sec. 4.7.3). Finally, it demands systematic inclusion of overlap in the calculations. The last difficulty is of no real importance, for it involves only some complications from the numerical point of view; the others, on the contrary, are serious ones, and place a limitation on the method.

Equation (3.4.2) also indicates that an essential difference of Hoffmann's method with respect to the standard Hückel method consists in the systematic use of the non-orthogonal basis formed by the *s*, *p*, ... orbitals of the various atoms.

3.4.1. Justification and limitations of EHT. - The method of Hoffmann has enjoyed great popularity, first in quantum chemistry, then in solid state physics. Its attractiveness stems from the fact that it retains the features that make the Hückel method an extremely efficient, if rough, scientific tool, and in addition introduces, even just for π systems, an explicit dependence of the hopping integrals on interatomic distances - thus eliminating the need to know which are the chemically linked atoms. A justification of the method is provided by (a) the remark that a second-quantization Hamiltonian can be associated with it, thus providing a model Hamiltonian:

$$\hat{H}_{EHT} = \hat{H}_0 + \sum_{s} \sum_{\mu,\nu} \frac{k}{2} \alpha_{\mu} \left(\hat{a}^+_{\mu s} S_{\mu\nu} \hat{a}_{\nu s} + \hat{a}^+_{\nu s} S_{\mu\nu} \hat{a}_{\mu s} \right) +$$

$$+ \sum_{s} \sum_{\mu} (1 - k) \alpha_{\mu} \hat{a}^+_{\mu s} \hat{a}_{\mu s} \qquad (3.4.3)$$

(b) the consideration of Durand's pseudopotential scheme, which gives for the off-diagonal terms of the pseudopotential a standard form, which, under Mulliken's approximation (3.4.2), amounts to

$$\sum_{A} < \mu | \hat{W}_{\mu,A} | \mu > = \frac{1}{2} S_{\mu\nu} \left(\sum_{A} < \mu | \hat{W}_{ps,A} | \mu > + \sum_{A} < \nu | \hat{W}_{ps,A} | \nu > \right) \qquad (3.4.4)$$

and, if M is the centre of $|\mu>$,

$$\sum_{A} < \mu | \hat{W}_{ps,A} | \mu > = \sum_{A \neq M} < \mu | \hat{W}_{ps,A} | \mu > + < \mu | \hat{W}_{ps,M} | \mu > \qquad (3.4.5)$$

Now, keeping Eq. (3.3.4) in mind, one can expect that the terms of the summation in (3.4.5) contain products of the squares of overlap integrals over different centres, whereas the last term is a one-centre term, and thus is likely to be much larger. Moreover, the first summation will have as its largest part a standard contribution from the (standard) surrounding of M in ordinary molecules. Therefore, (3.4.5) can be taken quite reasonably as an atomic quantity, possibly with changes depending on the coordination of the given atom (but cf. sec. 4.7.3). Of course, a more satisfactory parametrization could be obtained by direct use on the non-local potentials of Nicolas and Durand (1979).

Two limitations are evident in EHT: (i) the basis dependence of the parametrization and (ii) the absence of feedback effects for large charge shifts. (i) The former difficulty has been mentioned in the introduction of this section and in sec. 4.5: its consequences are mainly that no distinction can be made between the so-called "non-bonded interactions" and parameters associated with bonds. To a certain extent, this is an advantage, for it implies that the existence of bonds between specific pairs of atoms should be an output of the scheme. However, the geometry dependence of non-bonded interactions is greatly different from that of energy contributions associated with bonds, and therefore a simple scheme treating all the pairs of atomic orbitals on an equal footing is open to doubts, especially as regards conformational studies; such a defect would be mitigated if the method actually used a hybridized basis, whereby overlap integrals would be strongly dependent on whether two orbitals form a bond or not. This difficulty is even worse with methods like CNDO (sec. 4.5) which involve a parametrization independent of the angular parts of atomic orbitals.

(ii) The absence of feedback effects limits the reliability of the EH method in its original (non-SCF) form to small charge shifts. The reason is that the parameters of atoms which are close to negative or positive ions when *in situ* must be quite different from those of the same atoms when they are approximately neutral in the molecule. In addition to the obvious consequence that predicted electron shifts may be too large, serious difficulties in the comparison of molecules where atoms of the same species have different electron populations must be expected.

3.4.2. Applications of the EH method. - Extensive application of the EH method to all sorts of molecules can be found in the literature. An excellent survey is still the original paper by Hoffmann (1963). That method has also been widely applied to study clusters of atoms simulating solid-state problems (Messmer and Watkins 1973; Blyholder and Coulson 1967, 1968; Baetzold 1971, 1973). We illustrate one such application by the chemisorption study of Muda and Hanawa (1973). Those authors correctly chose for the α's the arithmetic averages of ionization potentials and electron affinities, a much more sensible choice than just the ionization potentials as used by many others (cf. secs. 4.4.1 and 4.7.3). The cluster studied is a nine-atom region of two close-lying (111) planes of diamond (the (111) crystal planes of diamond are alternately at distances of 1.542 and 0.514 Å). The nine carbon atoms correspond to the carbon skeleton of the quasi-planar conformation of 1.3.5-trimethylcyclohexane. A CO molecule is assumed to hover vertically above the centre of the cyclohexane ring. A number of interesting results are obtained by this simple model study. At 1.55 A from the surface, the 9C-CO configuration is $\sim$ 1.7 times more stable than the 9C-OC configuration, in agreement with chemical intuition. At the same distance (which is practically the CC bond distance) the oxygen atom of the

9C-CO complex appears to carry a maximum of electron population. As regards the nature of the levels, the authors compare the results for the cluster with free CO and with the band obtained by smoothing out the levels of the 9C cluster. They obtain a deep lying orbital ("split off state" at $\sim$ - 57 eV) which contains "the bonding combination of the $2s$ and $2p$ orbitals of the O atom, the $2s$ orbital of the C atom, and the bonding combinations of the $2s$ and the $2p$ orbitals of the $C_1, \ldots, C_6$ atoms directing to the C atom," and lies below the lower band edge. They conclude that the diamond-CO bond is essentially localized and is described mainly by the split-off state. The EH method has also been applied to CO on Ni and Cu (Itoh 1977).

Although quite concise, the study of Muda and Hanawa is a good example of both the advantages and the pitfalls of EHT (and similar methods). The main advantage is that it provides a quick and qualitatively correct picture of situations where the classical valency theory does not apply. The pitfalls depend on two reasons: (i) incorrect estimate of parameters; (ii) neglect of essential terms (with respect to rigorous expressions). Point (i) would apply, in the present instance, to the estimates of orbital-energy shifts upon formation of the 9C-CO bonds; point (ii) applies to the evaluation of binding energies, which should not be computed as sums of orbital energies as is done in many Hückel calculations, but should contain the effective nuclear repulsion term R_0' (sec. 4.2), especially when situations with different distances are compared (Del Re et al. 1977).

Concerning more specifically cluster calculations intended to simulate solid-state problems, the most delicate question is extrapolation to the solid. Both the number of atoms and the special selection of the model cluster are important. For example, the 9C cluster discussed above is chemically excellent for adsorption studies, but it simulates rather a diamond plane than a diamond three-dimensional lattice.

To the same line of work as that of Muda and Hanawa belong other applications of the EH method to chemisorption. Little known yet highly interesting papers on that topic are scattered here and there. For instance, in 1973 Moffat studied the adsorption or boron of H_2 with its axis parallel to the boron surface. The EHT method has also been used extensively by H. Müller and his school in Jena; in a series of papers they have presented models of several processes involving surfaces, trying to treat whole chemical processes like transport reactions at surfaces. One of the most fascinating applications is to the "balloon effect", which consists essentially of the removal of (gaseous) FeCO from a Fe surface subsequent to chemisorption of CO (Müller and Opitz 1976).

3.4.3. Systems with several heteroatoms: the parameter question. - Whereas in solid-state problems it is normally easy to discuss complications deriving from the choice of parameters, serious difficulties arise when complicated molecules with different atomic species are considered, both in the Hückel and in the EH method. It should be kept in mind that even overlap integrals are to some extent empirical parameters, due to their dependence on assumed bond distances and orbital exponents.

The parameter problem has been taken up several times in the present chapter. It will also be a point frequently considered in the next chapters. As has been already pointed out, the choice of parameters should obey a criterion of physical significance and of simplicity, if necessary at the expense of quantitative accuracy.

However, even with that criterion, difficulties in the assessment of results arise due to uncertainties in the choice of parameters such as those connected with overlap integrals. The resulting effects on the calculations are very difficult to analyze because of the large number of uncertain quantities. Moseley, Ladik and Mårtensson (1967) treated this problem on fairly complicated molecules - uracil and the four thiadiazole isomers. They worked within the framework of the Hückel method, but their procedure and results hold also for the EH method. The basic idea consists in using random numbers to generate Hamiltonian matrices containing random variations of the elements up to a fixed value (± 0.10 to $\pm 0.20\ \beta$) with respect to standard parameters. The results up to 150 tests are analyzed statistically in terms of averages and distribution moments.

The computations of Moseley *et al.* led to the following general conclusions:

a) The bond-order matrix appears to be quite stable with respect to random variations of the parameters;

b) The orbital energies oscillate much more, and therefore conclusions based on them are much less reliable.

The experience of many a quantum chemist tends to support both the above conclusions. Indeed, they are in line with what one would expect from perturbation theory. Nevertheless, (b) is often forgotten; whereas in fact contradictory results are obtained in the analysis of one-electron energy levels of molecules by different methods, as has been illustrated by Heilbronner and Schmelzer (1975; cf. sec. 4.5.3). There seems to be no way out of this, as even in *ab initio* schemes orbital energies are extremely sensitive to the basis and to the geometry.

Once again, the best suggestion is extreme care in defining the model used and the corresponding parameters, so that trends may be associated with situations where everything, including bond distances and orbital exponents, is given standard (ideal) values.

4.1. The SCF Hamiltonian.

As has been mentioned, a system of independent electrons is described by a single state of the n-electron basis *viz.* by an antisymmetrized product of appropriate spin-orbitals. For such a state to be an eigenstate of $\hat{H}_{\text{mod}}$ (Hamiltonian which replaces $\hat{H}_{\text{el}}$) the matrix representing $\hat{H}_{\text{mod}}$ in that basis must be diagonal:

$$<\Phi|\hat{H}_{\text{mod}}|\Phi'> \equiv <n_{1\uparrow},n_{1\downarrow}|\hat{H}_{\text{mod}}|n'_{1\uparrow},n'_{1\downarrow}\ldots> = \\ = E(n_{1\uparrow},n_{1\downarrow}\ldots)\,\delta(n_{1\uparrow},n'_{1\uparrow})\,\delta(n_{1\downarrow},n'_{1\downarrow})\ldots \qquad (4.1.1)$$

where $\delta(n,n')$ is Kronecker's symbol. It is easy to prove that (4.1.1) is satisfied if: (i) the basis states are orthogonal to each other, (ii) the two-electron part of $\hat{H}_{\text{mod}}$ vanishes, and (iii) the basis spin orbitals are eigenstates of the one-electron Hamiltonians whose sum gives $\hat{H}_{\text{mod}}$. Condition (iii) contains condition (i). In fact, the matrix elements of a Hamiltonian of the general form (2.2.1) depend on the scalar products

$$\lambda_{\text{js,ks}} = H^0_{\text{jk}} <\Phi|\hat{a}^+_{\text{js}}\hat{a}_{\text{ks}}|\Phi'> \qquad (4.1.2)$$

This is zero if $\hat{a}^+_{\text{js}}\hat{a}_{\text{ks}}$ does not take the right-hand state into the left-hand one and if the basis spin orbitals form an orthogonal set. If $\mathbf{H}^0$ is a diagonal matrix, then the only non-vanishing $\lambda_{\text{js,ks}}$ are those with $js = ks$, which means that the two states must be equal for $\hat{H}_{\text{mod}}$ to have a non-vanishing matrix element.

Terms in $\hat{H}_{\text{mod}}$ depending on four indices will be in general different from zero, because the spin orbitals that diagonalize $\mathbf{H}^0$ are unique; therefore, their disappearance must be imposed as a feature of the model. This is less restrictive than it might seem at first sight, because the one-electron part of $\hat{H}_{\text{el}}$ is not uniquely defined, and various one-electron effective Hamiltonians can be constructed.

4.1.1. A simplified form of the total Hamiltonian. - Consider the product

$$\hat{x} = \hat{a}^+_{\text{js}}\hat{a}^+_{\text{ls'}}\hat{a}_{\text{ms'}}\hat{a}_{\text{ks}} \qquad (4.1.1)$$

and the cases

a: $js = ls'$ b: $js = ms'$ c: $js = ks$

d: $ls' = ms'$ e: $ls' = ks$ f: $ms' = ks$

By the commutation rules (2.1.8) we have

$$\begin{array}{ll} \text{a: } \hat{x} = 0 & \text{b: } \hat{x} = \hat{a}_{ls}\, \hat{n}_{js}\, \hat{a}_{ks}\, \delta(s,s') \\ \text{c: } \hat{x} = \hat{a}^{+}_{ls'}\, \hat{n}_{js}\, \hat{a}_{ms} & \text{d: } \hat{x} = \hat{a}^{+}_{js}\, \hat{n}_{ls}\, \hat{a}_{ks} \\ \text{e: } \hat{x} = -\hat{a}^{+}_{js}\, \hat{n}_{ls}\, \hat{a}_{ms}\, \delta(s,s') & \text{B} = \hat{x} = 0 \end{array} \tag{4.1.2}$$

This allows us to write (changing names where necessary to dummy subscripts)

$$\begin{aligned} &\frac{1}{2}\sum (jk|lm)\hat{a}^{+}_{js}\, \hat{a}^{+}_{ls'}\, \hat{a}_{ms'}\, \hat{a}_{ks} = \\ &\frac{1}{2}\sum{}'''' (jk|lm)\, \hat{a}^{+}_{js}\, \hat{a}^{+}_{ls'}\, \hat{a}_{ms'}\, \hat{a}_{ks} + \\ &+ \sum [(jj|kl) - (jk|lj)\,\delta(s,s')]\hat{n}_{js'}\, \hat{a}^{+}_{ls}\, \hat{a}_{ks} \\ &- \frac{1}{2}\sum [(jj|ll) - (jl|lj)\,\delta(s,s')]\hat{n}_{js'}\, \hat{n}_{ls} \end{aligned} \tag{4.1.3}$$

where the quadruple prime means "all subscripts different". The second summation on the right hand side is obtained by applying (4.1.2), imposing that the subscripts be (l, j, j, k), and taking into account that $(lj|jk) = (jk|lj)$, $(jj|lk) = (lk|jj)$; the last summation is introduced to compensate for the fact that terms with $l = m$ and $j = k$ appear only once in the left-hand summation for a given j,l pair. Equation (4.1.3) allows us to write Eq. (2.1.14) in the new form

$$\begin{aligned} \hat{H}_{el} &= R_0 - \frac{1}{2}\sum_{js,ls} [(jj|ll) - (jl|lj)\,\delta(s,s')]\hat{n}_{js}\, \hat{n}_{ls'} + \\ &+ \sum_{js,ks} \hat{H}^{(s)}_{jk}\, \hat{a}^{+}_{js}\, \hat{a}_{ks} + \hat{J}_{corr} \end{aligned} \tag{4.1.4}$$

where J_{corr} is the first term in the right-hand side of (4.1.3) and

$$\hat{H}^{(s)}_{jk} = H_{jk} + \sum [(ll|jk) - (lk|jl)\,\delta(s,s')]\hat{n}_{ls'} \tag{4.1.5}$$

is the (j,k) element of a matrix $\hat{\mathbf{H}}^{(s)}$ representing *formally* a one-electron operator $\hat{H}^{(s)}_{1}$.

$\hat{\hat{H}}$ is not really a one-electron operator because of the presence of $\hat{n}_{ls'}$; its matrix elements are still operators, as is typical of two-electron operators. The steps in the construction of a general one-electron scheme are thus

(i) neglect of $\hat{J}_{corr}$

(ii) transformation of $\hat{\hat{H}}$ into an effective one-electron operator $\hat{H}_1^{eff}$ through two kinds of operations:

(ii-1) replacement of $\hat{n}_{ls'}$ by some constant;

(ii-2) specification of the basis orbitals.

4.1.2. The basis and the operator $\hat{n}_{ls}$. In order to proceed further it is important to remember that the spatial parts $|\underline{k}>$ of spin orbital bases $|\underline{k}\uparrow>$, $|\underline{k}\downarrow>$ on which the C.A. operators act is obtained in general from a given basis $|\underline{\mu}>$ by a linear transformation $\mathbf{T}$ obeying the orthonormality condition

$$\mathbf{T}^{+}\mathbf{S}\mathbf{T} = \mathbf{1} \tag{4.1.6}$$

just as $\mathbf{C}$ of Eq. (2.3.16), but not being necessarily the eigenvector matrix of a Hamiltonian matrix. Any element $|ks>$ of $|\underline{ks}>$ is given by

$$|ks> = (|\underline{\mu}\, s> \mathbf{T})_{ks} \equiv \sum |\mu s><\mu|k> \tag{4.1.7a}$$

and, conversely,

$$|\mu s> = \left(|\underline{k}s>\mathbf{T}^{-1}\right)_{\mu s} \equiv \sum |ks><k|\mu> \tag{4.1.7b}$$

where

$$<\mu|k> \equiv T_{\mu k} \qquad\qquad <k|\mu> \equiv (\mathbf{T}^{-1})_{k\mu} \tag{4.1.8}$$

We have, if $\hat{\theta}$ is any operator,

$$<j|\hat{\theta}|k> = [<\underline{k}|\hat{\theta}|\underline{k}>]_{jk} = [\mathbf{T}^{+}<\underline{\mu}|\hat{\theta}|\underline{\mu}>\mathbf{T}]_{jk} = \tag{4.1.9}$$

$$= \sum_{\mu\nu} <\mu|j>^{*}<\nu|k><\mu|\hat{\theta}|\nu>$$

and, in particular

$$(jk|ll) = \left\{\mathbf{T}^{+}<\underline{\mu}\left|\frac{ll}{r_{12}}\right|\underline{\mu}>\mathbf{T}\right\}_{jk} = \tag{4.1.10}$$

$$= \left\{\mathbf{T}^{+}<\underline{\mu}|\,(\mathbf{T}^{t}<\underline{\mu}\left|\frac{1}{r_{12}}\right|\underline{\mu}> \mathbf{T})_{ll}|\underline{\mu}>\mathbf{T}\right\}_{jk} = \sum T^{*}_{\mu j}T^{*}_{\nu l}T_{\rho l}T_{\tau k}(\mu\tau|\nu\rho)$$

$$(jl|lk) = \left\{ \mathbf{T}^{+} < \underline{\mu} \left| \frac{lk}{r_{12}} \right| \underline{\mu} > \mathbf{T} \right\}_{jl} = \tag{4.1.11}$$

$$= \sum T^{*}_{\mu j} T^{*}_{\nu l} T_{\rho k} T_{\tau l} (\mu\tau|\nu\rho)$$

with the current notation for two-electron integrals.

We have now

$$\sum_{ls'} [(jk|ll) - (jl|lk)\delta(s,s')]\hat{n}_{ls'}|\Phi> = \tag{4.1.12}$$

$$= \sum_{l} [(jk|ll)n_{l} - (jl|lk)n_{ls}]|\Phi> = J^{(s)}_{jk}|\Phi>$$

where n_l is the number of electrons (2,1,0) occupying the orbital l counted only once, and n_{ls} is 1 if l appears in an occupied spin orbital with spin s, zero otherwise. Then, by equations (4.1.10) and (4.1.11),

$$J^{(s)}_{jk} = \sum_{\mu,\nu} T^{*}_{\mu j} T_{\nu k} \sum_{\rho,\tau,l} T^{*}_{\rho l} T_{\tau l} \left[n_l(\mu\nu|\rho\tau) - n_{ls}(\mu\tau|\rho\nu) \right] \tag{4.1.13}$$

Following a well-known current practice, we also introduce the 'exchange' operator $\hat{\mathfrak{X}}$ defined by

$$<\mu|<\rho\left|\frac{\hat{\mathfrak{X}}}{r_{12}}\right|\tau>|\nu> = <\mu\tau|\rho\nu> \tag{4.1.14}$$

Then, the matrix whose jk-th element is given by (4.1.13) can be written:

$$\mathbf{J}^{(s)} = \mathbf{T}^{+} < \underline{\mu} | \hat{v}^{sh,s} | \underline{\mu} > \mathbf{T}, \quad \hat{v}^{sh,s} = \sum_{\rho,\tau} v^{(s)}_{\rho\tau}, \tag{4.1.15}$$

with

$$v^{(s)}_{\rho\tau} = <\rho\left|\frac{\hat{P}^{(s)}_{\rho\tau}}{r_{12}}\right|\tau>, \text{ with } \hat{P}^{(s)}_{\rho\tau} = \sum_{l} T^{*}_{\rho l} T_{\tau l}(n_l - n_{ls}\hat{\mathfrak{X}}) \tag{4.1.16}$$

4.1.3. <u>One-electron SCF Hamiltonian.</u> - Equation (4.1.12) is rigorous, and shows that the presence of $\hat{n}_{ls}$ in the expression of $\hat{H}_1$ involves a dependence on the state $|\Phi>$ under consideration; this is the characteristic point of all SCF-type one-electron models.

Let us now treat equation (4.1.12) as a recipe for replacing the operator part of (4.1.5) by a simple multiplier. Then we can write (4.1.5) in the form

$$\bar{H}^{(s)}_{jk} = H_{jk} + (\mathbf{T}^{+}<\underline{\mu}|\,\hat{V}^{sh,s}|\underline{\mu}>\mathbf{T})_{jk}, \tag{4.1.17}$$

or, with reference to the $|\underline{\mu}>$ (AO) basis:

$$\bar{\mathbf{H}}^{(s)} = \mathbf{H} + \mathbf{J}^{(s)}, \text{ with } \mathbf{J}^{(s)} = <\underline{\mu}|\,\bar{V}^{sh,s}|\underline{\mu}> \tag{4.1.18}$$

If **H** is written as the sum of a kinetic-energy contribution **K** plus a potential-energy contribution $\mathbf{V}^{core}$, Eq. (4.1.18) becomes

$$\bar{\mathbf{H}}^{(s)} = \mathbf{K} + \mathbf{V}^{eff,s} \text{ with } \mathbf{V}^{eff,s} = \mathbf{V}^{core} + \mathbf{J}^{(s)}. \tag{4.1.19}$$

Thus, $\mathbf{H}^{(s)}$ corresponds to a situation where the electron moves in a field which depends on the state under consideration through the occupation numbers and the coefficients $T_{\rho 1}$; it is the well-known SCF Hamiltonian. The SCF orbitals are obtained (for a given occupation scheme) by finding the **T** matrix which diagonalizes the matrix $\bar{\mathbf{H}}^{(s)}$ constructed with the same **T** in equation (4.1.16). Difficulties connected with the dependence on the spin s are normally avoided because only closed shells ($n_{1s} = n_1/2$) are taken into consideration.

4.2. Construction of Non-SCF Hamiltonians

Before turning to approximate SCF methods we discuss the foundations of non-SCF one-electron schemes in terms of Eq.s (4.1.5) and (4.1.18).

4.2.1. General expressions. - The first step toward construction of an effective one-electron Hamiltonian $\hat{H}^{eff}$ can be taken by replacing the operator $n_1 - n_{1s}\hat{x}$ of Eq. (4.1.16) by some sort of average over various states. To do that it is convenient to apply physical considerations. Formally, however, a statement like the following is sufficient:

> The contribution of electron-electron repulsion and exchange to the effective Hamiltonian is simulated by assigning a well specified set of effective occupation numbers to the molecular orbitals of a standard set, obtained from $|\underline{\mu}>$ by a suitable linear transformation **L**, and neglecting the exchange term of (4.1.13).

The above statement means that we replace $\hat{P}^{(s)}_{\rho\tau}$ of Eq. (4.1.16) by

$$P_{\rho\tau} = \sum_{l} L^{*}_{\rho l} n_{l} L_{\tau l}, \tag{1.2.1}$$

where the n's are the effective occupation numbers. As the latter are given as well as **L**, Eq. (4.1.18) is replaced by

$$\mathbf{H}^{\text{eff}}(\underline{\mu}) = \mathbf{H}(\underline{\mu}) + \mathbf{J}'(\underline{\mu}) \tag{4.2.2}$$

where the argument $\underline{\mu}$ just reminds that the matrices are referred to the AO basis. The Hamiltonian (4.2.2) is a Hückel-like Hamiltonian.

More generally, one may include in the model Hamiltonian only part of the effects present in **H**. This is for instance the case when non-bonded interactions are to be treated as separate effects. Then, instead of **H**, we write an **H'** which differs from it in some specified way:

$$\mathbf{H}^{\text{eff}}(\underline{\mu}) = \mathbf{H}'(\underline{\mu}) + \mathbf{J}'(\underline{\mu}). \tag{4.2.3}$$

Equations (4.2.1) and (4.2.3) define a general non-SCF one-electron Hamiltonian over an AO basis. The corresponding many-electron operator is, of course

$$\hat{H}^{\text{eff}} = R'_{0} + \sum_{js,ks} H^{\text{eff}}_{jk} \hat{a}^{+}_{js} \hat{a}_{ks} \tag{4.2.4}$$

where $\mathbf{H}^{\text{eff}}$ is now referred to a basis $|\underline{k}>$ obtained from $|\underline{\mu}>$ by a linear transformation **T**. If **T** diagonalizes $\mathbf{H}^{\text{eff}}(\underline{\mu})$, then

$$\hat{H}^{\text{eff}} = R'_{0} + \sum_{j,s} \varepsilon_{jj} \hat{n}_{js}, \tag{4.2.5}$$

so that the energy associated with a state obtained by occupying in a given way the elements of $|\underline{k}>$ is just the constant term plus the sum of the orbital energies of the occupied spin orbitals. This is a characteristic of Hückel-like models, which does not mean, however, that comparison of different molecules is allowed directly in terms of orbital energies, because of the term R'_0. In fact, inclusion of R'_0 is the simplest way to take into account the general remarks made in sec. 4.2.5.

The preceding results can be discussed with reference to five points:

(i) the Hückel model
(ii) determination of R'_0
(iii) choice of the basis $|\underline{\mu}>$
(iv) effects neglected in $\mathbf{H}'$
(v) form of $\mathbf{J}'$ and of $\mathbf{H}^{\text{eff}}$

4.2.2. The Hückel model. - As can be shown from Eq. (4.2.1) - and was shown by a different formalism already a long time ago (Del Re and Parr 1963) - , the Hückel model involves simplifying assumptions on the effective potential and on the apparent nuclear charges of the atoms participating in the given molecular system. In practice, the operator $\hat{V}^{\text{sh},s}$ is assumed to provide the potential corresponding to a uniform distribution of the $n - 1$ electrons seen by a given electron over the various atoms of the molecule or π system under study. The resulting effective core is strongly positive only when heteroatoms sharing a lone pair are present; for instance, the π system of the pyrrole ring C_4H_4NH will be equivalent to four charges equal to the core charge of the carbon atoms decreased by 1 (and therefore approximately zero) and to one charge corresponding to the core charge of nitrogen (approximately + 2) also decreased by 1; the nitrogen atom will create roughly the potential of a univalent positive ion. A first step toward the Hückel model consists in choosing the effective potential according to the effective core just illustrated, *i.e.* in replacing the actual system and its SCF Hamiltonian by an electron moving in a model external field.

The Hückel model also implies approximations like the nearest-neighbour one, and, at least in its original form, the orthogonality condition for the atomic-orbital basis. Such approximations also correspond to a model treatment, but they are partly connected with the choice of the basis, which will be discussed later. Here we only consider the matrix elements of the Hamiltonian, and the n.n.a. The latter amounts to claiming that

$$\beta^{(\text{Hü})}_{\mu\nu} \approx K_{\mu\nu} + <\mu|\sum_{\rho,\tau} V^{\text{eff},s}_{\rho\tau}|\nu> = 0 \text{ for } \mu,\nu \text{ not neighbours} \qquad (4.2.6)$$

(obtained by taking an off-diagonal element of (4.1.19) over the AO (Hückel) basis). We have already assumed that β is actually obtained from the potential of an effective core; here we have the additional requirement that the potential in question is not necessarily a Coulomb potential, but one which, for a suitable set of atomic orbitals, satisfies (4.2.6).

The above statement illustrates the difficulties encountered when a noncontradictory model is desired. In fact, if it is possible in general to determine a potential satisfying (4.2.6) for any given set of point charges and atomic orbitals, it will not be possible in general to obtain the same β for the same bond in different molecules. Thus, the very possibility of considering the Hückel method as a model treatment hinges on a delicate mathematical point, if the n.n.a. is made an essential part of it. We shall assume here that for standard π systems the condition (4.2.6) can be satisfied by suitable orthogonalized orbitals; but we emphasize that the question is not settled, and may be important for certain applications, *e.g.* to solids.

The other question is whether the model reflects reality. Of course, this question can only be answered by indirect evidence. Such evidence is partly to be found in agreement with experiment, but, as has been mentioned, there are serious risks in requiring simplified models to agree accurately with experiment. Additional indications can come from numerical studies; for instance, the SCF PPP model seems to lead to an effective Hamiltonian matrix which roughly satisfies the n.n.a. (Carpentieri, Porro, Del Re, 1968).

4.2.3. Determination of R_0'. - The zero-point energy R_0' corresponds to the repulsion of the 'dressed' cores, as can be deduced from equation (4.1.4). Using twice the approximation (4.2.1), one obtains

$$R_0' = R_0 - \frac{1}{2}\sum P_{\rho\tau}P_{\sigma\nu}(\rho\tau|\sigma\nu) \tag{4.2.7}$$

The physical meaning of this expression is simple; the nuclear or core repulsions appearing in R_0 are shielded by the attractions of pairs of densities $P_{\rho\tau}(\rho\tau)$ and $P_{\sigma\nu}(\sigma\nu)$, $(\rho\tau)$ and $(\sigma\nu)$ being AO products. The origin of this correction can be explained as follows. Suppose that it is possible to obtain a good approximation to $(\rho\tau|\sigma\nu)$ by the generalized Mulliken approximation:

$$\begin{aligned}(\rho\tau|\sigma\nu) \cong \frac{1}{4}f_{\rho\tau}f_{\sigma\nu}(\rho\rho|\sigma\sigma) + \frac{1}{4}f_{\tau\rho}f_{\sigma\nu}(\tau\tau|\sigma\sigma) + \\ + \frac{1}{4}f_{\rho\tau}f_{\nu\sigma}(\rho\rho|\sigma\sigma) + \frac{1}{4}f_{\tau\rho}f_{\nu\sigma}(\tau\tau|\nu\nu)\end{aligned} \tag{4.2.8}$$

where $\mathbf{f}$ is an *ad-hoc* matrix, equivalent to a generalized overlap.

Then

$$\frac{1}{2}\sum P_{\rho\tau}P_{\sigma\nu}(\rho\tau|\sigma\nu) \cong \frac{1}{8}\sum_{\rho,\sigma}\left(\sum_{\tau} P_{\rho\tau}f_{\rho\tau}\right)\left(\sum_{\nu} P_{\sigma\nu}f_{\sigma\nu}\right)\cdot$$

$$\cdot\ (\rho\rho|\sigma\sigma) + \ldots = \frac{1}{2}\sum_{\rho,\sigma} Q_\rho Q_\sigma(\rho\rho|\sigma\sigma) = \tag{4.2.9}$$

$$= \sum_{\rho<\sigma} Q_\rho Q_\sigma(\rho\rho|\sigma\sigma) + \frac{1}{2}\sum_{\rho} Q_\rho^2\ (\rho\rho|\rho\rho),$$

the last equality being justified most easily by assuming that **L** of (4.2.1) is a real transformation, so that $P_{\rho\tau} = P_{\tau\rho}$, and by keeping in mind that in a summation over ρ,τ,σ,ν the subscripts can be interchanged at will. The atomic orbital population Q_ρ is defined here as

$$Q_\rho = (\mathbf{fP})_{\rho\rho}. \tag{4.2.10}$$

Now consider two cores having positive charges Z_A, Z_B and carrying electrons providing a shielding equivalent to charges Q_A, Q_B, respectively. The total mutual electrostatic energy of A and B can be written:

$$\frac{(Z_A - Q_A)(Z_B - Q_B)}{R_{AB}} = \frac{Z_A Z_B}{R_{AB}} - \frac{Q_A Q_B}{R_{AB}} - \frac{Q_A(Z_B - Q_B)}{R_{AB}} - \frac{Q_B(Z_A - Q_A)}{R_{AB}} \tag{4.2.11}$$

The latter pair of terms in the right hand side represents the potential energy of the electrons at A in the field of the core + electrons of B, and vice versa; the first two terms represent the rest of the potential energy. Now, the electron-core attractions are taken into account in the one-electron Hamiltonian of Eq. (4.2.4); R_0' must contain the core repulsion *minus* the electron cloud repulsions; and the latter are precisely what Eq. (4.2.9) represents, including a self-energy term.

We conclude that the constant term R_0' in a model Hamiltonian of the type (4.2.4) is a fundamental one for comparisons of energies of different molecules: Even in the Hückel method it is not legitimate just to compare sums of orbital energies. Arguments to the contrary have been given especially in connection with cluster calculations *e.g.* by Messmer and Watkins (1973). Such arguments are acceptable as long as they refer to closely similar geometrical situations, but they become invalid in two cases: (a) if the energies of different systems have to be compared, because then the zero-point of vibrational energy must be common to all of them; (b) if the change in energy upon change in geometry is studied. These points are extremely important for cluster computations, *e.g.* when models of inhomogeneous alloys are studied; but they are best illustrated by molecular calculations. For instance, both experiment and theory suggest that the first UV transition of cyclopentadiene is shifted to the red with respect to butadiene. In a study of possible energy transfer, say, between butadiene and cyclopentadiene, it is important to know whether that red shift is due to a lowering of the first excited state or to a raising of the ground state of cyclopentadiene with respect to butadiene. The indications tend towards the latter alternative, as is also suggested by chemical intuition (Del Re, Rastelli, Momicchioli, 1968).

4.2.4. Analysis of $\mathbf{H}^{eff}$; the AO basis. - The matrix $\mathbf{H}^{eff}$ of Eq. (4.2.3) represents a linear one-electron Hamiltonian; one which, unlike its SCF equivalent, does not depend on the states on which it acts. This has been obtained by assigning to all the electrons but one the fixed distribution (4.2.1). This is the physical picture underlying the Hückel-like methods.

However, the assumption about the effective field is only one of two basic assumptions of those methods. The other one regards the basis. We have assumed that the reference basis $|\underline{\mu}>$ is in practice one of atomic orbitals, even though care has been taken not to imply that it is necessarily so. The obvious reason why it is interesting to refer to atomic orbitals is that we want atoms to figure explicitly in our formulas. More precisely, following the octet rule, we want one atomic orbital per valency of any first-row atom; this allows us to think of at most eight electrons surrounding that atom, and of two-electron bonds formed by coupling of one atomic orbital per atom. A basis so defined will be called a MVAO basis.

Thus, the matrix $\mathbf{H}^{eff}$ is a finite one over a relatively simple basis. We shall refer later to the conceptual difficulties involved in the definition of the MVAO basis (Ch. 5).

Also the basis $|\underline{k}>$ may be required to have special characteristics, which will necessarily imply special assumptions on $\mathbf{H}^{eff}$. Such characteristics reduce basically to block factorization of $\mathbf{H}^{eff}$. The simplest example is the assumption implicit in the standard Hückel method for π-systems that a separate block of $\mathbf{H}^{eff}$ need only be treated to describe π-electron states. (An illustration of the σ-π separation on the example of CO_2 is given in Chapter 5).

In the case of the Hückel method the MVAO basis is just supposed to satisfy the σ-π separation, but in many other cases actual construction of suitable molecular-orbital bases is required prior to formulation of a model Hamiltonian. For instance, the Newns-Anderson Hamiltonian of sec. 4.3.2 is defined on the crystal orbitals of the substrate - *i.e.*, the basis is supposed to diagonalize the one-electron Hamiltonian matrix of the substrate. In general, the approximations contained in a model Hamiltonian are not invariant under a linear transformation of the basis, and the basis used for a given model must be carefully specified.

4.2.5. SCF Methods as Independent Particle Models. - The analysis of the Hartree-Fock theory in terms of a model has originated ardent controversies between some physicists and some chemists concerning the very nature of the various SCF independent-particle models (SCF-IPM). For simplicity, consider a closed-shell electron system described by a single Slater determinant $|\psi>$. In the 'operatorial interpretation', attention is focused on the fact that the total Hamiltonian is approximated by a sum over the N electrons of the system, namely

$$\hat{H} \simeq \sum_{\nu=1}^{N} \hat{F}(\nu) \tag{4.2.12}$$

where $\hat{F}(\nu)$ is the standard one-electron effective Hamiltonian of the Fock-Roothaan method:

$$\hat{F} = \hat{H}^{\text{core}} + \sum_{k} (\hat{J}_k - \tfrac{1}{2}\hat{K}_k)\hat{n}_k \tag{4.2.13}$$

n_k being the occupation number of the orbital $|\varphi_k>$ in $|\psi>$ ($n_k = 2$ or 0 in the closed shell case). Equations (4.2.12) and (4.2.13) have led some people, especially solid-state physicists, to consider the SCF approximation as a true physical IPM. Accordingly, they try to interpret the whole experimental reality in terms of one-electron orbital energies (ε_{kk}) and one-electron contributions (a_k) associated to a given observable; the former are connected to the $\hat{F}$ operators by diagonalization procedures (e.g. the energy bands), and the latter are obtained when breaking up the mean value of an observable written in the form of a sum of one-electron operators $\hat{A}$ into orbital contributions.

Such an interpretation appears to be supported by the fact that, in the SCF approximation, knowledge of the first-order density matrix implies knowledge of the higher-order density-matrices, as was pointed out in well-known work by Dirac, Fock, Löwdin, McWeeny, and others; a result which is self-evident, because it is simply related to the one-electron form of the corresponding Dirac operator. That attitude is adopted in quantum chemistry when people speak of Walsh rules, frontier orbitals, Woodward-Hoffmann rules, and more generally when an orbital picture of the electronic structure of atoms and molecules is given.

In the 'energetic interpretation', attention is focused on the form taken by the energy E of an electron system in Coulombic interaction, if it is calculated as the mean value of the total exact Hamiltonian

$$<\psi|\hat{H}_{\text{exact}}|\psi> = \sum_{k} n_k I_k + \sum_{j,k} \frac{n_j n_k}{2} (J_{jk} - \tfrac{1}{2} K_{jk}) \tag{4.2.15}$$

On several occasions, Mulliken has emphasized that this expression cannot be reduced to the sum of the energies ε_{kk} of the electrons occupying the orbitals $|k>$ in $|\psi>$; by partitioning it into its orbital components

$$I_k \text{ and } \varepsilon_{kk} = I_k + \sum_j n_j (J_{kj} - \frac{1}{2} K_{kj}),$$

he showed instead that the total energy may be written as combinations of various forms, e.g.

$$E = \frac{1}{2} \sum_k n_k (I_k + \varepsilon_{kk})$$

(additive partition). All SCF calculations, including the semi-empirical and the *ab initio* ones, use the complete form of E in the problems of molecular conformations and related topics. The same holds for problems involving the energy difference between two different states, for instance the transitions energies. An apparent exception is the case of ionization potentials, studied in the context of the approximation of Koopman's theorem.

Recently, an effort has been made to reduce approximately the SCF total energy to a weighted sum of SCF orbital energies: this seems to corroborate the interpretation of the SCF-IPM as an IPM in the strictest sense. The starting point of these developments is an approximation derived by Ruendenberg (1977) from Politzer's considerations (1976) on the virial theorem. Unfortunately, that expression is not precise enough for quantitative predictions, in spite of several attempts to refine it by using *ad hoc* parameters instead of the theoretical factor 1.5.

A striking test of orbital-energy arguments has been given by Kertesz *et al.* (1978). Those authors carried out ab initio calculations on several extended systems in order to verify whether the exact SCF total energy can be approximated by a relation of the form

$$E \simeq 2(1 + x) E_0$$

where E_0 is a sum of orbital energies ε_{kk}. Calculating the values of x for infinite chains, they found important variations of x, which goes from -0.023 (chain of H atoms) to about 0.45-0.50 (chains of carbon atoms and hydrocarbons) and to 0.69 (HF chain). They concluded that stability problems cannot be discussed in terms of a sum of orbital energies E_0, even in systems large enough to belong to the realm of solid-state physics.

4.3. Many-Electron Models and their SCF Versions

So far we have concentrated our attention on semi-empirical one-electron schemes of the simplest type. We shall now discuss more general forms of model Hamiltonians. Even if in practice some of them are reduced to one-electron schemes by an approximation of the Hartree-Fock type, they all start with some physical assumption on the two-electron terms of Eq. (2.1.14).

A classical model was proposed by Pariser and Parr for π systems, and extended to give the well-known CNDO model in its various versions. The essential approximation consists in assuming that only the so-called two-electron Coulomb integrals, *i.e.* those between electrons assigned each to a single spin-orbital, give a non-vanishing contribution. The explicit form of the resulting Hamiltonian will be given in sec. 4.5.1, after presentation of a few simpler versions of it.

The general Hamiltonian from which the SCF models are obtained do not apply, of course, just to one-electron schemes; it is perfectly possible to build a configuration interaction scheme on them, and in fact some methods like PCILO are conceived directly in terms of CI. The great advantage of CI calculations is that they allow treatment of excited states without any special caution. The difficulties are: a) the selection of important configurations; b) the interpretation of results.

At present, with the exception of PCILO and of the still developing multi-configurational SCF (SCF-MC) approaches, the popular quantum chemical methods are SCF-MO ones even at the cost of ignoring correlation energy (v. Herigonte 1972). This is due, among other things, to the fact that computer programmes are available and not yet fully exploited. There is little doubt, however, that sooner or later configuration interaction will be normally used to supplement orbital calculations.

The above remark should be kept in mind because the fact that the following sections are mostly devoted to SCF treatments could be misleading.

4.3.1. Hubbard and Anderson Hamiltonians. - The general Hamiltonian puts all the atomic pairs of basis spin orbitals on an equal footing, as is reasonable in an ordinary molecule. One might suppose that in fact only electrons occupying the *same* orbital contribute to the two-electron term. This implies a further simplification, and gives the Hubbard Hamiltonian (Hubbard 1955 a, b, 1957, 1958);

$$\hat{h}_{\text{Hub}} = h_0' + \sum_{\mu s} \alpha_\mu \hat{n}_{\mu s} + \sum_{\mu s, \nu s} \beta_{\mu\nu} \hat{a}^+_{\mu s} \hat{a}_{\nu s} + \sum_{\mu} \gamma_{\mu\mu} \hat{n}_{\mu\uparrow} \hat{n}_{\mu\downarrow} \qquad (4.3.1)$$

This type of Hamiltonian is much used in solid-state theory, and contains the idea that only localized electrons interact with one another. That idea, in turn, implies that a basis of atomic orbitals be used and that interactions between orbitals of the same atoms vanish or that there is just one orbital per atom. The Anderson Hamiltonian assumes that only one atomic orbital contributes to the two-electron term of (4.3.1), as will be shown below.

4.3.2. Interaction between an atom and a solid. - In order to introduce semiempirical SCF models, we shall refer to the very topical problem of chemisorption.

A very simple model of chemisorption is the Newns (1969) model for the adsorption of a one-electron atom (hydrogen) on a metal. The binding energy between the adsorbed atom ('adatom') A and the substrate S, and the charge transfer to or from the adatom are the quantities characteristic of that phenomenon.

The model in question is analogous to that of the hydrogen molecule-ion; therefore it is based on the orbitals characteristic of the separated systems - the adatom orbital $|a\rangle$ and the metal Bloch orbitals $|\underline{k}\rangle$. Clearly, the question of MAO's arises also in the present context, but we shall consider the adaptation of the atom and of the metal to the situation a minor effect.

As regards electron-electron interactions, it can be safely assumed that one more electron in such an enormous system as the metal does not disturb the situation; whereas the electron repulsion associated with more than one electron occupying $|a\rangle$ is important. Therefore, the Anderson (1961) simplified form of the Hubbard Hamiltonian is used:

$$\hat{H}^{(\mathrm{ads})} = \varepsilon_a \hat{n}_{a\uparrow} + \varepsilon_a \hat{n}_{a\downarrow} + \sum_k \varepsilon_k \hat{n}_{k\uparrow} + \sum_k \varepsilon_k \hat{n}_{k\downarrow} + \sum_{k,s=\uparrow,\downarrow} (V^+_{ak} \hat{a}^+_{ks} \hat{a}_{as} + V_{ak} \hat{a}^+_{as} \hat{a}_{ks}) + \gamma_{aa} \hat{n}_{a\uparrow} \hat{n}_{a\downarrow}, \tag{4.3.2}$$

This Hamiltonian can be put into SCF form by substituting one of the $\hat{n}_a$'s with its expectation value and keeping only terms with the opposite spin:

$$\hat{H}_1^{(\mathrm{ads.}\uparrow)} \equiv \hat{H}_1^{(\mathrm{ads})} = (\varepsilon_a + \gamma_{aa} n_{a\downarrow})\hat{n}_{a\uparrow} + \sum_k \varepsilon_k \hat{n}_{k\uparrow} + \sum_k (V^*_{ak} \hat{a}^+_{k\uparrow} \hat{a}_{a\uparrow} + V_{ak} \hat{a}^+_{a\uparrow} \hat{a}_{k\uparrow}) \tag{4.3.3}$$

(and the corresponding one for spin $\downarrow$), where $n_{a\downarrow}$ is the electron population of the spin orbital $|a\downarrow\rangle$ ($n_{a\downarrow} \equiv \langle \hat{n}_{a\downarrow} \rangle$). The physical picture behind (4.3.3) is quite simple: the 'electronegativity' of the adatom adjusts to the electron transfer by a sort of linear response.

The interaction is between the orbital of the adatom A and the Bloch orbitals of the substrate S. Nevertheless, when one tries to obtain some kind of physical evaluation of the parameters, the atomic orbital picture is needed, because one wants to assume that A interacts only with surface atoms of S; this is how the standard MO-LCAO picture creeps in and the question of binding between atoms arises again.

The Anderson Hamiltonian as given in Eq. (4.3.3) corresponds to a matrix

$$\mathbf{H}^\uparrow = \begin{vmatrix} \varepsilon_\uparrow & V_{a1} & V_{ak} \cdots & V_{an} \\ V_{1a} & \varepsilon_1 & & 0 \\ \vdots & & \ddots & \\ V_{ka} & & \varepsilon_k & \\ \vdots & 0 & & \ddots \\ V_{na} & & & \varepsilon_n \end{vmatrix} \tag{4.3.4}$$

where $\varepsilon_\uparrow$ is obtained iteratively from

$$\varepsilon_\uparrow = \varepsilon_a + \gamma_{aa} n_{a\downarrow} \tag{4.3.5}$$

and $n_{a\downarrow}$ is obtained by solving simultaneously the secular equation for $\mathbf{H}^\downarrow$. The ε_k's are the crystal orbital energies.

The determination of $n_{a\downarrow}$ is made by the standard formula

$$n_{a\downarrow} = \sum_{m\downarrow\,\mathrm{occ}} |\langle a\downarrow | m\downarrow \rangle|^2 \tag{4.3.6}$$

where $|m\downarrow\rangle$ is the m-th molecular orbital of the combined system. A modified from of (4.3.6) is used when the crystal orbitals form a continuum.

4.3.3. The average occupation number. - Equation (4.3.6) contains two assumptions; the first one is that the best average electron population to be used in the SCF form of the Hubbard or Anderson Hamiltonian corresponds to the weight of $|a\downarrow\rangle$ in $|m\downarrow\rangle$; the other, that $|a\rangle$ is orthogonal to all the $|k\rangle$ orbitals.

The first assumption as such may be defended on the ground that we are considering the interaction of a spin-orbital with its counterpart built with the same spatial orbital. The second one shows here its implications. To suppose that $|a\rangle$ is orthogonal to every $|k\rangle$ means assuming that $|a\rangle$ derives from a standard atomic orbital $|a_{sk}\rangle$ through some pollution by the $|k\rangle$ orbitals (which form by definition an orthogonal set):

$$|a\rangle = \frac{1}{\sqrt{1 - |\langle k|a_{st}\rangle|^2}}\left[|a_{st}\rangle - \sum_k |k\rangle\langle k|a_{st}\rangle\right] \tag{4.3.7}$$

(The orthogonalization made here is *not* the Löwdin orthogonalization because we want Eq. (4.3.4) to retain its "reduced coupling" form. If the $|k\rangle$'s were made to take up some contribution of $|a_{st}\rangle$, the interaction elements between the new $|k\rangle$'s would not be zero.) Suppose now that $|a\rangle$ is given by (4.3.7); then there are two alternatives:

(i) it is not justified to assume that γ_{aa} of Eq. (4.3.3) is a characteristic of the adatom, because it is actually given by

$$\begin{aligned}\gamma_{aa} \equiv (aa|aa) = \frac{1}{N^4}\Big\{&(a_{st}a_{st}|a_{st}a_{st}) - 4\sum_k s_{ka}(ka_{st}|a_{st}a_{st}) + \\ &+ 2\sum_{k,k'} s_{ka}s_{k'a}[(kk'|a_{st}a_{st}) + 2(ka_{st}|k'a_{st})] + \\ &+ 4\sum_{k,k',k''} s_{ka}s_{k'a}s_{k''a}(kk'|k''a_{st}) + \sum_{k,k',k'',k'''} s_{ka}s_{k'a}s_{k''a}s_{k''a}(kk'|k''k''')\Big\}\end{aligned} \tag{4.3.8}$$

with $N^2 = (1 - \sum|\langle k|a_{st}\rangle^2)^2$, $s_{ka} \equiv \langle k|a_{st}\rangle$; even if all the two-electron integrals or their coefficients except the first one could be neglected, γ_{aa} *would depend on the particular molecular situation* because of the denominator.

(ii) $n_{a\downarrow}$ and $n_{a\uparrow}$ are so redefined that one can set

$$\gamma_{aa} \cong (a_{st}a_{st}|a_{st}a_{st}) \tag{4.3.9}$$

for instance, still with the assumption given above,

$$n_{a\downarrow} \doteq \frac{\sum |<a\downarrow|m\downarrow>|^2}{(1 - \sum |<k|a_{st}>|^2)} . \tag{4.3.10}$$

Of course, one can results the orthogonality assumption by claiming that the s_{ka} 's are anyway very small, so that (4.3.9) is valid and (4.3.10) coincides with (4.3.6); or one can reformulate the whole treatment with the inclusion of overlap, as has been done above. The former point of view has a certain attractiveness, for it might be argued that there are many cases where chemisorption bonds are not as strong as regular chemical bonds; in those cases the V_{ak} 's could be comparatively large for the equilibrium distances, the overlap integrals being negligible. The "weak chemisorption limit" implies that the bond integral between the adatom and the nearest surface atom or atoms is much smaller than between the metal atoms.

The above considerations illustrate many of the points raised in the previous chapters: in particular, the importance of internal consistency even of very simple methods and models for determining the physical problems to which they apply. The above considerations also imply a precise notion of what $|a_{st}>$ is: with an atomic orbital compatible with the ZDO approximation, the question of overlap would disappear altogether. The quest for the optimal MVAO's again appears to be of the utmost importance.

4.4. Methods with Iterative Determination of Atomic Parameters

4.4.1. In situ atomic orbital energy. - Another special feature of the SCF Hamiltonian (4.3.3) is the orbital energy (4.3.5). That expression, which we have obtained here from a model Hamiltonian proposed in 1961, was already suggested by Coulson and Rushbrooke in 1940, was used among others by A. Pullman and Berthier in 1949 (cf. sec. 4.5), and was discussed by Pople (1953) and Fischer-Hjalmars (1965: cf. Jug 1969, p. 112). In fact, it lies at the basis of the ω-techniques of Wheland and Mann (1949), Hinchliffe (1967), Gayoso (1971, 1972), Gayoso *et al.* (1974), and of Mathur *et al.* (1977). Extension of Eq. (4.3.5) to all the atoms of a molecule or of a cluster (from the Hubbard Hamiltonian) is straightforward. If we consider only non-magnetic states we can write

$$\varepsilon_{\mu}^{(val)} = \varepsilon_{\uparrow} = \varepsilon_{\downarrow} = \varepsilon_{\mu} + \frac{1}{2}(\mu\mu|\mu\mu)n_{\mu} \tag{4.4.1}$$

$|\mu>$ being an atomic orbital, and the superscript (val) denoting the atom *in situ*. The possibility of estimating $\varepsilon_{\mu}^{(val)}$ stems from the following interpretation of (4.4.1); the electron population n_{μ} of $|\mu>$ is equally divided into spin up and spin down, so that an electron in $|\mu>$ with spin up will feel (on the average) the repulsion of $n_{\mu}/2$ electrons in the same orbital with spin down, and vice versa; this is a valence-state orbital energy (cf. sec. 1.6.2). On the other hand, the ionization potential I_{μ} of the isolated atom when $|\mu>$ goes from singly-occupied to empty must coincide with the negative of ε_{μ}, for in that case we do assign to the single electron a given spin; when $|\mu>$ is doubly occupied, expression (4.4.1) must coincide with the negative of the electron affinity A_{μ}. Thus, we can write

$$-A_{\mu} = -I_{\mu} + (\mu\mu|\mu\mu) \tag{4.4.2}$$

and, as a result,

$$\begin{aligned} \varepsilon_{\mu}^{(val)} &= -I_{\mu} - \frac{1}{2}(A_{\mu} - I_{\mu})n_{\mu} = \\ &= -\frac{I_{\mu} + A_{\mu}}{2} + \frac{1}{2}(A_{\mu} - I_{\mu})q_{\mu} = \\ &= -\chi_{\mu}^{(Mu)} + \frac{1}{2}(A_{\mu} - I_{\mu})q_{\mu} \end{aligned} \tag{4.4.3}$$

where q_{μ} is the net charge in a.u. associated with n_{μ} (i.e. the negative of the excess electron population of $|\mu>$ whose standard population is assumed to be one) and $\chi^{(Mu)}$ is the Mulliken electronegativity already discussed in sec. 1.6. The critical remarks made in sec. 1.6.2 hold also in this case, of course; and it is also clear that Eq. (4.4.2) introduces a number of effects not included in the model through the estimate of the two-electron integral $(\mu\mu|\mu\mu)$ - obviously an empirical procedure which ought to be separately justified. As long as that is not done explicitly the arguments leading to (4.4.3) may be modified to provide other expressions of $\varepsilon^{(val)}$ in terms of atomic properties. The main difficulty is that the arguments used interpolate an expression which should refer to the atom *in situ* between limiting cases corresponding to the isolated atom. Further work in this connection seems to be needed.

4.4.2 Models based on equation (4.4.1). - If the self-consistency requirement is dropped, an average value of n_μ must be taken, and parameters suitable for methods like the extended Hückel theory are found. Alternatively, an SCF interpretation of the method for inductive effects described in sec. 3.2 can be obtained by assuming that (a) $n_{A(B)}$ - the population of the orbital $| A(B) >$ which A uses to form a bond with B - is a linear function of the difference $\varepsilon_{B(A)} - \varepsilon_{A(B)}$, and that (b) $\varepsilon_{A(B)}$ is an average value over the bonds formed by A, a value which, with a suitable unit and a suitable zero point, is just the δ_A parameter of Eq. (3.2.1). Then equation (4.4.1) becomes

$$\delta_{A(B)} = \delta^{oo}_{A(B)} + \frac{1}{2} J_{A(B)} \left[a + b\left(\delta_{B(A)} - \delta_{A(B)}\right)\right] \tag{4.4.4}$$

with $\delta^{oo}_{A(B)}$, a, b appropriate constants, and $J_{A(B)} \equiv (A(B)A(B)|A(B)A(B))$. This, when averaged over the N_A neighbours of A (and keeping in mind that also $\delta_{B(A)}$ must be replaced by an average) gives:

$$\delta_A \simeq \frac{1}{N_A} \sum_B \left[\delta^{oo}_{A(B)}\left(1 - \frac{b}{2} J_{A(B)}\right) + \frac{a}{2} J_{A(B)}\right] + \sum_B \frac{bJ_{A(B)}}{2N_A} \delta_B . \tag{4.4.5}$$

Comparison with (3.2.1) shows that (4.4.5) is actually equivalent to it. Further discussion of (4.4.5) is outside the scope of the present review.

The most current application of Eq. (4.4.1) is found in the ω-techniques mentioned above, which are π-electron methods especially suited for hydrocarbons, where the extra assumption is made that

$$(\mu\mu|\mu\mu) = 2\omega\beta \tag{4.4.6}$$

ω being a constant with a value of 1.4 (Streitwieser 1961). If β is a standard bond integral and the atomic orbitals differ only in their locations, it is clear that (4.4.6) is always correct, and the only role of ω is to provide a value for $(\mu\mu|\mu\mu)$. Therefore the ω-technique corresponds exactly to a π-electron method based on the Hubbard Hamiltonian.

The two-parameter ω-technique of Mathur *et al.* (1977) involves both a dependence of the type (4.4.1) for the diagonal elements of the Hamiltonian and a dependence of the bond-integrals on the bond orders. Thus, it is essentially an SCF procedure of the PPP type, improved to extend self-consistency to the bond integrals, much as has been done by Gayoso (1972).

4.5. PPP Method and its Extension (CNDO family).

4.5.1 PPP method. - The Hamiltonian associated with the well known PPP (Pariser-Parr-Pople) method (Parr and Pariser 1953, Pople 1953) was given explicitly only in 1967 by Koutecky. The latter pointed out that, given a basis of N "localized" (atomic) orthonormal spin orbitals $|\mu s\rangle$, the PPP Hamiltonian (using R_0 as a zero point) could be written:

$$\hat{H}^{PPP} = \sum_s \left\{ \left[\sum_\mu \alpha_\mu^{00} \hat{n}_{\mu s} + \sum_{\mu,\nu}{}' \beta_{\mu\nu}^0 \hat{a}_{\mu s}^+ \hat{a}_{\nu s} - \sum_{\mu,\nu}{}' J_{\mu\nu} Z_\nu \hat{n}_{\nu s} + \right.\right.$$
$$\left.\left. + \frac{1}{2} \sum_\mu J_{\mu\mu} Z_\mu \hat{n}_{\mu s} \right] + \frac{1}{2} \sum_\mu \hat{n}_{\mu s} \left(\sum_{\nu, s'} J_{\mu\nu} \hat{n}_{\nu s'} - J_{\mu\mu} \right) \right\}, \qquad (4.5.1)$$

where $J_{\mu\nu}$ stands for $(\mu\mu|\nu\nu)$ and Z_μ, Z_ν are core atomic charges per orbital. This expression corresponds to the form which the general Hamiltonian (2.1.14) takes when it is referred to orthonormal atomic orbitals for which

$$(\mu\rho|\nu\sigma) = \delta_{\mu\rho}\delta_{\nu\sigma} J_{\mu\nu} \qquad (4.5.2)$$

(which is either an ad hoc ansatz or a consequence of the more general zero-differential-overlap ansatz (5.7.1)).

The relationship of (4.5.1) with the simple Anderson and Hubbard Hamiltonians is immediately seen if it is written in the simpler form

$$\hat{H}^{PPP} = \sum_s \left[\sum_\mu \alpha_\mu^0 \hat{n}_{\mu s} + \sum_{\mu,\nu}{}' \beta_{\mu\nu}^0 \hat{a}_{\mu s}^+ \hat{a}_{\nu s} - \frac{1}{2} \sum_{\nu, s'}{}' J_{\mu\nu} \hat{n}_{\mu s} \hat{n}_{\nu s'} \right] \qquad (4.5.3)$$

with α_μ^0 given by

$$\alpha_\mu^0 = \alpha_\mu^{00} + (\mu\mu|\mu\mu)(Z_\mu - \tfrac{1}{2}) - \sum_\nu (\mu\mu|\nu\nu) Z_\nu \qquad (4.5.4)$$

If Z_μ can be identified as the number of electrons lost by $|\mu\rangle$ to the molecular electron cloud, it is evident that (4.5.4) coincides with (4.4.1) when $n_\mu = 1$ and the two-centre integrals $(\mu\mu|\nu\nu)$ are neglected.

The SCF form of (4.5.4) is found in the same way as with the Anderson Hamiltonian. The one-electron Hamiltonian for a given spin is

$$\hat{H}^{\uparrow}_{\mathrm{SCF}} = \sum_{\mu} \alpha^{(\uparrow)}_{\mu} \hat{n}_{\mu\uparrow} + \sum_{\mu,\nu}{}' \beta_{\mu\nu} \hat{a}^{+}_{\mu\uparrow} \hat{a}_{\nu\uparrow} , \qquad (4.5.5)$$

with

$$\alpha^{(\uparrow)}_{\mu} = \alpha^{0}_{\mu} + (\mu\mu|\mu\mu) n_{\mu\downarrow} + \frac{1}{2} \sum_{\nu,\mu}{}' \sum_{s'} (\mu\mu|\nu\nu) n_{\nu s'} . \qquad (4.5.6)$$

The history of the PPP method is a glorious one, because it provided the first SCF scheme for molecular calculations. It is now well known to everybody, and is incorporated into the all-valence-electron methods of the CNDO family (what we shall call the "NDO" methods). The book (and reprint collection) of Parr on that method (1963) is a classic. It is important to go back to those old papers to gain an understanding of the foundations of the NDO methods.

4.5.2. All-valence PPP (NDO)-methods. - The approximate SCF schemes considered so far are characterized by the fact that each atom carries one orbital (normally *s* in solids and π in molecules). The next steps in the generalization of models of the PPP type are:

a) extension to more than one orbital per atom;
b) improved formulas for the evaluation of parameters;
c) removal of orthogonality restrictions.

The first two steps were the central points of the well-known CNDO family, whose recent versions are mostly due to Dewar. We shall not stop on those methods because there are already excellent reviews (*e.g.* Klopman and O'Leary, 1970; Jaffé, 1969; Jug, 1969; Dewar, 1975). The various methods differ in the recipes used for approximating two-electron integrals and core parameters. For instance, the MINDO/3 scheme applies the zero-differential overlap approximation only to many-centre two-electron integrals, whereas one-centre integrals are estimated from spectroscopic data. The core parameters and the two-electron integrals are evaluated by *ad hoc* parametric formulae.

The whole scheme is quite reasonable even though it is not easy to justify theoretically the neglect of differential overlap (NDO). Indeed Dewar (1975) claimed that the method is free from the usual objections made against semiempirical methods, since it reproduces correctly most ground-state properties of a large number of molecules. Nevertheless, the CNDO family illustrates the difficulties that face semiempirical methods when two much emphasis is placed on quantitative agreement, so that the only interest is in accurate and quantitatively reliable predictions.

The idea of a Hamiltonian of the PPP type (4.5.1) holding for all the valence orbitals at the cost of a few very simple approximations (Pople *et al.*, 1965) is extremely interesting and physically significant. The use of intuitive physical arguments and simplifications for the parameters are also part of the qualities of the method. For instance, the assumption that $(\mu\mu|\nu\nu)$ is independent of the angular parts of $|\mu>$ and $|\nu>$ is a simplification which can be understood very easily as meaning that orbitals are replaced by spherical *s* like distributions in the model underlying the method. Similar considerations hold if the Dewar-Sabelli (1962) split-orbital recipe is adopted. The point where difficulties arise is when the method is tested against experimental results with the simple criterion: Disagreement = wrong, agreement = right. Attempts to improve the parameters according to such a criterion may reduce the physical significance of the method, unless it is clearly indicated how the physical model changes.

In an excellent short review, Jaffé (1969) gave a number of examples of the changes in agreement-disagreement patterns of CNDO/1 and CNDO/2; the latter is a version of the former with a more sophisticated choice of parameters partly based just on computational experience - a procedure whereby the basic physical assumptions of the model are modified in an unknown way.

Similar remarks hold for the whole CNDO family, as is mentioned at the end of Sec. 4.5.5. A more constructive way of looking at things can be illustrated on the example of two-electron integrals. Suppose that we have reason to believe that certain properties, not correctly predicted by the CNDO model (*i.e.* the CNDO scheme with a choice of parameters uniquely associated with given physical assumptions), are related to electron repulsions. Now, in CNDO the latter are averaged over the angular parts of atomic orbitals; this implies that directional effects are not included in the model. Therefore, what the model tells is the extent to which the properties in question are independent of the directional properties of electron repulsion; it is neither wrong nor right, it is simply incomplete.

4.5.3 Applications of the NDO methods: Hoffmann-Heilbronner "trough-space" and "through-bond" interactions. - The methods under consideration have been perhaps the most popular semiempirical methods for molecules and clusters in the last decade. Whereas little or no use of them has been made to construct interpretational schemes of chemical facts, an immense number of papers has been concerned with numerical results. A critical study, regarding in particular geometries and thermodynamical data with comparison of different NDO and *ab-initio* methods, has been given by M. C. Flanigan *et al.* (1977), a review with a conclusion containing a convincing argument in favour of semiempirical methods.

Consistent with the spirit of the present review, we shall consider other aspects of the applications of NDO methods, in particular their use for the physical interpretation of chemical effects.

As has been pointed out several times, the simpler a method the easier its use as a reference model for analyzing the interplay of effects in a given molecule. Indeed, simple qualitative rules are often most powerful tools of research (Dewar 1971). Hoffmann's work has been largely inspired by that consideration; in addition to the famous Woodward-Hoffmann rules, it has produced a large number of valuable ideas along these lines. The concepts of "through-space" and "through-bond" interactions is one of them (Hoffmann 1971). As a matter of fact, those concepts can be associated with a simple orbital scheme such as EHT; however, an assessment of them at the SCF level is of the highest importance. That task was performed by Heilbronner and Schmelzer (1976). We shall use their work as an illustration of the way in which interpretational problems can be handled at the SCF level, and of an application to organic chemistry of methods of the CNDO family.

The notion of a classification of interactions comes quite spontaneously from the consideration that two atomic orbitals $|\mu\rangle$, $|\nu\rangle$ can be coupled to each other either directly or through a third atomic orbital $|\rho\rangle$ (or both). The two limiting cases are represented in the Hückel or EHT scheme by $\langle\mu|H_{Hü}|\nu\rangle \equiv \beta_{\mu\nu} \neq 0$ and $\beta_{\mu\nu} = 0$, $\beta_{\mu\rho}$ and/or $\beta_{\nu\rho} \neq 0$ respectively.

In both cases $|\mu\rangle$ and $|\nu\rangle$ can be treated as forming a pair of bond orbitals; in the former case this is the result of direct interaction, in the latter it results from the removal of the degeneracy of the bonding-antibonding pair $(|\mu\rangle \pm |\nu\rangle)/\sqrt{2}$ due to coupling with $|\rho\rangle$. In other words, the two cases correspond to a non-vanishing first-order perturbation term and to a second-order perturbation effect, respectively. Note that, when a molecule is symmetric, the above considerations should refer to symmetry-adapted combinations of bond orbitals.

The experimental analysis of a molecular system in terms of orbital energies can be carried out with the help of photo-electron spectroscopy; but, of course, the assignment of experimentally detected one-electron energies is a problem for theoreticians. Much of Heilbronner's recent work (Bieri *et al.* 1976, 1977; Batich *et al.* 1974; Bloch *et al.* 1978; Bischof *et al.* 1978; Allan *et al.* 1975) is devoted to that problem. The importance of introducing concepts like "through-space" and "through-bond" interactions derives from the fact that these effects can shift the energies of given orbitals enough to affect their order, and hence the corresponding assignment. For instance, the lone-pair orbitals $|n_1\rangle$ $|n_2\rangle$ of the molecule formed by two nitrogen atoms connected by three $-CH_2-CH_2-$bridges (Hoffmann *et al.* 1970) can be considered to form a pair of non-degenerate bonding-antibonding orbitals as a result of through-bond interactions, with obvious consequences on the ordering of the various levels.

Although, as has been mentioned, the 'natural' framework in which the effects under consideration can be defined is a Hückel-type scheme (because there no feedback effects come into play) a redefinition of them in the framework of an SCF scheme is possible, as Heilbronner (1975) has given along the following lines. First of all, the matrix elements of the SCF Hamiltonian are referred to a basis of localized orbitals (LMO) chosen according to a well specified localization scheme. Next, these LMO's are combined to form SLMO's, *i.e.* a basis for the irreducible representations of the symmetry group of the given molecule. Finally, the n SLMO's $|\varphi,k\rangle$ associated with a given irreducible representation are considered; and, one of them being of interest at any time, the other $n-1$ SLMO's of the same irreducible representation are replaced by the eigenvectors of the corresponding $(n-1) \times (n-1)$ SCF Hamiltonian matrix. The final orbitals are called by Heilbronner 'precanonical' and give rise to a transformed Fock matrix $\mathbf{F}$ (s,k) which differs from $\mathbf{F}$ by (i) being in general an $n \times n$ block smaller than $\mathbf{F}$, because associated with the s-th irreducible representation $\Gamma^{(s)}$ (ii) being represented in a basis such that only the k-th line and column contain off-diagonal elements different from zero; in other words, a matrix of the form (4.3.4) up to the order of the rows and columns.

Through-space interaction is now measured by the element $F_{\lambda_1\lambda_2}$ of $\mathbf{F}^{(LMO)}$ which couples two localized orbitals $|\lambda_1\rangle$, $|\lambda_2\rangle$; through-bond interaction is measured by the coupling elements in $\mathbf{F}^{(r,b)}$ and/or in $\mathbf{F}^{(s,a)}$ between the bonding and antibonding combinations $|\rho_b^{(r)}\rangle$ and $|\rho_a^{(s)}\rangle$ of $|\lambda_2\rangle$, respectively, and other precanonical orbitals of the appropriate sets. (Of course, if the differences in the diagonal elements of the perturbing orbitals are large, in both cases the 'coupling' is not measured by the corresponding off-diagonal element, but by its square divided by that difference). Note that localized orbitals are not in general atomic orbitals; in fact, although the brief description given above referred explicitly to AO's, it holds also for other localized orbitals. At any rate, in cases like non-bonding lone-pair orbitals, the localized orbitals are very close to atomic orbitals.

With the above scheme Heilbronner has compared the predictions of several CNDO-like methods: CNDO/2 (Dewar and Haselbach, 1970), SPINDO (Asbrink *et al.*, 1972) for three classes of molecules representing different limiting cases, norbornadiene as a specific application, 1.4-cyclohexadiene and its bridged derivatives in connection with the effects of changes in dihedral angle, and the three-bridge molecule mentioned above in connection with interaction between lone pairs.

The extensive numerical work and the detailed analysis of Heilbronner cannot be reported here. But we report his rather pessimistic conclusion concerning the validity of the three methods: "We can 'justify' a preconceived idea about the relative importance of through-space vs. through-bond interaction by 'objectively' performing a calculation, using an appropriately chosen standard-SCF procedure without tampering with its original parametrization!"

4.5.4 Applications of NDO methods to polymers and crystals. - Two types of applications of the PPP and CNDO family to periodic polymers and crystals are current: cluster calculations and calculations with periodic boundary conditions. The former have been extensively performed by researchers like Baetzold (1976). There are many delicate aspects in such calculations, for they are often performed using standard programmes and accepting the results at their face-value. It would be interesting if comparative studies like that of Heilbronner and Schmelzer were carried out also in cluster studies.

The formulation of SCF schemes in periodic systems is by no means a matter of routine. The reason is that such systems must be treated by block-factorization of the total Hamiltonian matrix and by diagonalizing the resulting blocks (Ladik 1965 a,b; Del Re *et al.* 1967); therefore, the iterations required by an SCF procedure demand special care. Nevertheless, a systematic formulation of the application to such systems of the various NDO methods, starting with the pure-π PPP method, was produced by Ladik (1965 b) and by Ladik and Biczó (1971 a,b). A very interesting aspect of those studies is the systematic allowance for spin-dependent solutions of the SCF equations, a possibility implicit in the fact that the Hamiltonian (4.1.5) may be used to define SCF orbitals $|j\rangle$, $|l\rangle$ which depend on the spin s (unrestricted SCF).

As an illustration of the type of calculations that can be carried out, we mention work on DNA models as well as saturated and unsaturated hydrocarbon chains. Protein-model studies have suggested in particular that the role of next-nearest neighbours is very important in the band structure of proteins (Ladik, 1974). Study of polyacetylene, polyethylene, and polyglycine (Beveridge *et al.*, 1972) and DNA (Suhai and Ladik, 1973) provide extensive comparisons between PPP, CNDO/2, INDO, MINDO/2 (and EHT) band structures. The MINDO/2 method appears to give more reasonable bands (for say, polyglycine) than the INDO method - in accordance with the fact that that method seems to produce better ionization potentials.

Finally, the case of DNA models, where an extensive π-system is essentially responsible for the bands, gives the interesting result that MINDO/2 reproduces the results of PPP π-calculations, whereas CNDO/2 gives band widths larger by 3-4 eV than the other two methods.

In general, the pessimistic conclusion quoted from Heilbronner in the preceding section could be extended to these band calculations.

As concerns conformations - another topic of primary interest for polymers and biopolymers, but less important for the crystals usually considered by solid state physicists - a great deal of work has been done by a relatively restricted number of researchers, others preferring strictly empirical treatments possibly supplemented by intramolecular charge-transfer effects estimated from simple net-charge considerations (e.g. the Madelung-energy considerations of Jørgensen, 1967). Extensive quantum studies are due to B. Pullman (1976 and references therein).

4.5.5 Critique of NDO procedures. - The remarks by Heilbronner reported at the end of Sec. 4.5.3 are not directed against simplified and/or semiempirical methods as such. Indeed, they lend further support to an idea which is the guiding principle of the present study; semiempirical methods are respectable mainly as models, and too much effort to make them 'work' may destroy their value. One could claim, for instance, that what Heilbronner calls a 'preconceived idea' could acquire a measure of scientific validity if it were a 'feature of the model' just like the independence of energies from bond angles in the Hückel method, due to the nearest-neighbour approximation. The difficult point is that parameters are not always evaluated according to clearcut physical assumptions.

For example, suppose that the zero-differential overlap (ZDO) approximation were interpreted as an assumption on the basis AO's; then it would mean that the off-diagonal matrix elements of *any* one-electron operator over those basis orbitals should vanish. Therefore, either the ZDO approximation for two-electron integrals is equivalent to some other model feature which is not incompatible with assuming non vanishing off-diagonal elements of one-electron operators, or the current parametrizations introduce inconsistencies in the models. It will be shown in section 4.6 that the inconsistencies deriving from the ZDO approximation can be removed in a complete way. We do not claim, however, that this is sufficient to remove all inconsistencies of SCF methods deriving from the original PPP scheme.

The interest of viewing methods just as interpolation formulas was strongly advocated by Dewar (1975; Dewar and Haselbach 1970). As has been mentioned, there are reasons not to discard that type of 'philosophy'. The difficulty is that it ends up in a sequence of newer and newer general methods. The consequences of wishing to obtain an entirely general interpolation scheme are shown by the paper of Heilbronner mentioned above, and by the studies of Ladik on the band structures of polymers; but other examples exist. In addition to the comments of Jaffé on CNDO, it is known that CNDO/2 does not produce good rotation barriers (Gropen and Seip, 1971; Perahia and A. Pullman 1973) ; the MINDO/3 method does not produce good results for unusual bonding, like the double bond in $H_2Si = CH_2$ (Bantle and Ahlrichs 1978).

Many objections and conceptual difficulties could be removed if simple models were defined and simplifications introduced at the basis level (choice of the trial function) and at the level of effective core potentials.

4.6. A General version of All-Valence SCF Methods: the SCF Extended Hückel Method. (BMV Method)

From the practical point of view as well as in principle, a consistent and complete definition of the parameters of any molecular orbital method can only be given if the exact structure of the effective Hamiltonian $\hat{H}^{eff}$ which governs electron motion is known. We have already described Durand's analysis of this question for the Hückel method. A similar procedure consists in treating $\hat{H}_1^{eff}$ as a Hartree-Fock Hamiltonian by letting

$$\hat{H}_1^{eff} = \hat{H}_1^{core} + \sum_j d_j(\hat{J}_j - \frac{1}{2}\hat{K}_j) \qquad (4.6.1)$$

where d_j is the number (2,1,0) of the electrons which occupy $|j\rangle$ (the j-th molecular orbital) in the state under consideration. This expression, which introduces the Coulomb and exchange operators ($\hat{J}_j$ and $\hat{K}_j$) associated with $|j\rangle$ (Roothaan, 1951) already mentioned in sec.s 4.1.2 and 4.1.3, amounts to the Fock Hamiltonian of the current closed-shell SCF theory and to the approximate Hamiltonian of Longuet-Higgins and Pople (1955) for free radicals, triplet states, etc.

4.6.1 Expressions of two-electron integrals. - Rigorous computation of the matrix elements $H_{\mu\mu}^{eff} = \alpha_\mu$ and $H_{\mu\nu}^{eff} = \beta_{\mu\nu}$ for a given basis of m orbitals requires m^2 core integrals $H_{\mu\nu}^{core}$ and m^4 two-electron integrals $(\mu\nu|\rho\tau)$, which represent the two-electron interactions.

As has been recalled in sec. 4.3, most semi-empirical methods can be made to derive from Eq. (2.1.14), the differences being in the choice of the parameters α_μ, $\beta_{\mu\nu}$, and $(\mu\nu|\rho\tau)$. For instance the CNDO-INDO method of Pople *et al.* (1965, 1967) mentioned in sec. 4.8 rests (among other things) on the zero-differential-overlap (ZDO) approximation, which amounts to assuming that the basis orbitals are orthogonal and two-electron integrals in which at least one of the electrons is associated with different basis orbitals vanish. A more general method is provided by the SCF extended-Hückel method of Berthier, Millié, and Veillard (1965), which works with a non-orthogonal basis and adopts for two-electron integrals the Mulliken approximation already mentioned in Sec. 3.4, supplemented by Ruedenberg's (1951) expression:

$$(\mu\mu|\nu\rho) = \frac{1}{2} < \nu|\rho > [(\mu\mu|\rho\rho) + (\mu\mu|\nu\nu)], \tag{4.6.2a}$$

$$(\mu\nu|\rho\mu) = \frac{1}{2} < \nu|\rho > [(\mu\rho|\rho\mu) + (\mu\nu|\nu\mu)]. \tag{4.6.2b}$$

These formulas make it possible, at variance from the methods of the ZDO family, to take all the two-electron integrals into account, even though they are expressed in terms of overlap integrals $<\mu|\rho>$ and of two-centre Coulomb and exchange integrals.

4.6.2 Core Hamiltonian matrix. - In the SCF-EHT method, only the valence electrons of the various atoms of a molecule are considered, and their states are determined variationally in terms of the minimal valence AO basis (one $1s$ AO for H, one $2s$ and three $2p$ AO's for C, N, O, five $3d$, one $4s$ and three $4p$ AO's for Fe, Co, Ni, etc.). The core Hamiltonian $\hat{H}_1^{\text{core}}$, which includes, in addition to the kinetic energy, the potential energy of the given valence electron with respect to the nuclei and to the electrons of the various atoms can be written in the pseudopotential form

$$\hat{H}_1^{\text{core}} = \hat{T}_1 + \sum_{\text{atoms L}} \hat{V}_L \tag{4.6.3}$$

Following a suggestion by Goeppert-Mayer and Sklar (1938), generalized by Moffitt (1951) in his method of 'atoms in molecules', the theory of valence states (cf. sec. 1.6.2) is used to produce an explicit approximate form of the core potentials $\hat{V}_L$. If the atom L participates with v_L valence electrons in the system under study, $\hat{V}_L$ is taken as the Hartree-Fock potential of the v_L-th positive ion, measured with respect to the potential $\hat{V}_L^0$ of neutral L:

$$\hat{V}_L = \hat{V}_L^0 - \sum_{\lambda \in L} n_\lambda^L \left(\hat{J}_{\lambda\lambda}^L - \frac{1}{2} \hat{K}_{\lambda\lambda}^L \right) \tag{4.6.4}$$

n_λ^L being the population of the AO $|\lambda>$ of L in a valence state chosen according to the electronic structure of the molecule; for instance, the V_4 $sxyz$ state of carbon, or even states where certain orbitals are hybridized.

4.6.3 Calculation of matrix elements of $\hat{H}^{eff}$. - Consider a matrix element $H_{\mu\mu} = \langle\mu|\hat{H}_1^{eff}|\mu\rangle$ corresponding to an atomic orbital $|\mu\rangle$ of the atom M. It is convenient to write

$$\hat{H}_1^{core} = \hat{T}_1 + \hat{V}_M + \sum_{L \neq M} \hat{V}_L \tag{4.6.5}$$

so that

$$H_{\mu\mu}^{core} \equiv \langle\mu|\hat{H}_1^{core}|\mu\rangle = W_\mu^M + \sum_{L \neq M} \langle\mu|\hat{V}_L|\mu\rangle \tag{4.6.6}$$

Here W_μ^M is a purely atomic quantity whose formal expression is obtained by comparing the energies of M neutral and v_M times ionized in appropriate valence states, with the same set of AO's in both cases; in other words, conditions are chosen such that the Koopmans' approximation for the evaluation of ionization potentials I^M holds. In practice, it is convenient to introduce a term $H_{\mu\mu}^0$ that can be identified to an ionization potential or to a neutral atom electronegativity, and the two-electron integrals corresponding to the interaction of the given electron with the other valence electrons of M are subtracted from it (Tables 4.6.1 and 4.6.2). These purely atomic terms can be extracted from atomic spectroscopy tables in the same way as Tables 1.2 and 1.3; they are known combinations of Slater-Condon parameters. The same line of reasoning can be applied to the summation over L in Eq. (4.6.6), and therefore

$$\langle\mu|\hat{H}_1^{core}|\mu\rangle = W_\mu^M + \sum_{L \neq M} \Big\{ \langle\mu|\hat{V}_L^0|\mu\rangle + \\ - \sum_{\lambda \in L} n_\lambda^L \Big[(\lambda\lambda|\mu\mu) - \tfrac{1}{2}(\lambda\mu|\mu\lambda) \Big] \Big\} \tag{4.6.7}$$

where $\hat{V}_L^0$ is a neutral-atom potential, and $\langle\mu|\hat{V}_L^0|\mu\rangle$ is a 'penetration integral' (in the sense of π-electron theories of the PPP type, but changed in sign). These are short-range two-centre terms, just like the exchange terms, and decrease exponentially with the distance R_{ML} between the nuclei M and L; therefore, both exchange and penetration integrals can be neglected in (4.6.7) with respect to the long-range integrals of the Coulomb type, which behave asymptotically as $1/R_{ML}$.

The interaction terms associated with valence electrons and contained in the Hamiltonian $\hat{H}_{\text{eff}}$ give rise to diagonal matrix elements of the form

$$G_{\mu\mu} = \langle\mu| \sum_j \left(\hat{J}_j - \frac{1}{2}\hat{K}_j\right)|\mu\rangle = \sum_j d_j \sum_{\rho,\tau} c^*_{\rho j} c_{\tau j}\left[(\mu\mu|\rho\tau) - \frac{1}{2}(\mu\tau|\rho\mu)\right] \tag{4.6.8}$$

where the sum over ρ,τ extends to all the basis AO's and $c_{\tau j}$ is the coefficient of $|\tau\rangle$ in $|j\rangle$. If the Mulliken-Ruedenberg approximation (4.6.2) is now applied to the two-electron integrals appearing in (4.6.8), the latter expression becomes

$$G_{\mu\mu} = \sum_\rho \left[(\mu\mu|\rho\rho) - \frac{1}{2}(\mu\rho|\rho\mu)\right] P_{\rho\rho} \tag{4.6.9}$$

where (in the case of real coefficients)

$$P_{\rho\rho} = \sum_j \sum_\tau d_j c_{\rho j} c_{\tau j} S_{\rho\tau} \tag{4.6.10}$$

is the so-called gross population of $|\rho\rangle$ (Mulliken 1955).

Expression (4.6.9) contains only one-centre and two-centre two-electron integrals over AO's $|\nu\rangle$ belonging to M (the centre of $|\mu\rangle$) as well as to other centres. With this remark in mind, the expression of α_μ can be written

$$\alpha_\mu \equiv H^{\text{eff}}_{\mu\mu} = H^{\text{core}}_{\mu\mu} + G_{\mu\mu} = W^{\text{M}}_\mu + \sum_{\nu\in \text{M}} P^{\text{M}}_{\nu\nu}\left[(\mu\mu|\nu\nu) - \frac{1}{2}(\mu\nu|\nu\mu)\right] + \sum_{\text{L}\neq \text{M}}\left\{\langle\mu|\hat{V}^0_{\text{L}}|\mu\rangle - \sum_{\lambda\in \text{L}} q^{\text{L}}_\lambda\left[(\mu\mu|\lambda\lambda) - \frac{1}{2}(\mu\lambda|\lambda\mu)\right]\right\} \tag{4.6.11}$$

with

$$q^{\text{L}}_\lambda = n^{\text{L}}_\lambda - P_{\lambda\lambda} \tag{4.6.12}$$

the 'net charge' associated with $|\lambda\rangle$.

Neglecting the penetration integrals on the ground that they correspond to short-range forces and the term containing the net charges q^{L}_λ on the ground that the introduction of atomic valence states implies that in most compounds the latter are very small, equation (4.6.11) reduces to

$$\alpha_\mu \approx W_\mu^M + \sum_{\nu \in M} P_{\nu\nu}^M \left[(\mu\mu|\nu\nu) - \frac{1}{2} (\mu\nu|\nu\mu) \right] \tag{4.6.13}$$

This expression is a generalization of the expression (4.4.1) which has been derived from the Hubbard Hamiltonian and used to justify the ω-techniques, and of (4.5.6), the corresponding expression for the PPP methods. All the data needed to compute the α's explicitly can be obtained from atomic spectroscopy data, since (as has been mentioned) the two-electron integrals $(\mu\mu|\nu\nu)$ and $(\mu\nu|\nu\mu)$ reduce to combinations of Slater-Condon parameters. Thus, for light atoms:

$$(ss|ss) = (ss|xx) = F^0 \qquad (sx|xs) = \frac{1}{3} G^1$$

$$(xx|xx) = F^0 + \frac{4}{25} F^2$$

$$(xx|yy) = F^0 - \frac{2}{25} F^2 \qquad (xy|yx) = \frac{3}{25} F^2 \tag{4.6.14}$$

Tables (4.6.1) and (4.6.2) contain the numerical values of these parameters for light atoms (Berthier *et al.* 1965 b); for metals using d orbitals, the corresponding values for Fe, Co, Ni, Cr (De Brouckère 1970) as well as tables allowing recalculation for the transition metals of the first two periods are available (di Sipio *et al.* 1971). It is evident that the simplified formula (4.6.13) is only acceptable if intermolecular charge transfers are weak. In the case of highly polar compounds, q^L is greatly different from n^L; it is no longer possible to suppress the last term in the complete expression of α_μ, but at least its long-range part must be retained. In fact, if those integrals are reinserted in the expression of α_μ in the asymptotic form (inverses of internuclear distances), the Madelung potential created by intramolecular charge transfers is superposed to the SCF potential of the *in situ* atom M, in accordance with ideas repeatedly expressed by Jørgensen (1967). In practice, it is better not to use the asymptotic expression in its naive form, because the values it gives do not tend to the one-centre integrals for a vanishing internuclear distance R_{ML}, but interpolation formulas of the Mataga-Nishimoto (1956) type

$$(\mu\mu|\nu\nu) = \left[R_{ML} + \frac{2}{(\mu\mu|\mu\mu) + (\nu\nu|\nu\nu)} \right]^{-1} . \text{ (in a.u.)} \tag{4.6.15}$$

appear to be quite satisfactory (de Brouckère, 1978).

With regard to the off-diagonal elements $H_{\mu\nu}$, the approximation used by most authors in the standard EHT (sec. 3.4) was the WH formula (3.4.1). As was mentioned, the kinetic energy elements $T_{\mu\nu}$ are not well represented by (3.4.2) - which corresponds to (4.6.2a) - whereas it is satisfactory for potential energy terms (Newton *et al.*, 1966). A more reasonable procedure then consists in actually computing the kinetic energy integrals $T_{\mu\nu}$ and $T_{\mu\mu}$, and setting

$$H_{\mu\nu} = T_{\mu\nu} + \frac{1}{2} S_{\mu\nu} \left[\left(H_{\mu\mu} - T_{\mu\mu} \right) + \left(H_{\nu\nu} - T_{\nu\nu} \right) \right] \tag{4.6.16}$$

4.6.4. Model Hamiltonian. - With the approximations (4.6.2) a many-electron Hamiltonian can be written for the method under consideration in the form:

$$\begin{aligned} \hat{H}_{BMV} = R_0 + \sum H^{core}_{\mu\nu} \hat{a}^+_{\mu s} \hat{a}_{\nu s} + \\ + \sum_{\mu,s} \sum_{\rho,s'} \left[(\mu\mu|\rho\rho) - (\mu\rho|\rho\mu)\delta_{ss'} \right] \hat{n}_{\rho s'} \hat{m}_{\mu s} \end{aligned} \tag{4.6.17}$$

where

$$\hat{m}_{\mu s} = \hat{a}^+_{\mu s} \hat{a}_{\mu s} \tag{4.6.18a}$$

$$\hat{n}_{\rho s'} = \frac{1}{2} \sum_{\tau} \left[\hat{a}^+_{\tau s'} <\tau|\rho> \hat{a}_{\rho s'} + \hat{a}^+_{\rho s'} <\rho|\tau> \hat{a}_{\tau s'} \right] \tag{4.6.18b}$$

and the two-electron terms over four different AO's have been neglected - in virtue of the fact that they are of the order of squares of overlap integrals $<\mu|\nu>$. Insertion of the WH approximation for the potential energy part of $\mathbf{H}^{core}$ gives

$$\begin{aligned} \hat{H}_{BMV} = R_0 + \sum_{\mu,\nu,s}{}' T_{\mu\nu} \hat{a}^+_{\mu s} \hat{a}_{\nu s} + \sum_{\mu,s} V^{core}_{\mu\mu} (\hat{n}_{\mu s} - \hat{m}_{\mu s}) + \\ + \sum_{\mu,s} \left\{ H^{core}_{\mu\mu} + \sum_{\rho,s'} \left[(\mu\mu|\rho\rho) - (\mu\rho|\rho\mu)\delta_{ss'} \right] \hat{n}_{\rho s'} \right\} \hat{m}_{\mu s} \end{aligned} \tag{4.6.19}$$

where $\mathbf{H}^{core}$ is given by (4.6.7). The SCF form of $\hat{H}_{BMV}$ is obtained by replacing $\hat{n}_{\rho s'}$ by the gross atomic population.

Two remarks should be added:

(a) In open-shell systems (e.g. free radicals) as well as in solid-state computations, spin-unrestricted (or magnetic) states can be studied simply (cf. Anderson 1961) by writing one SCF Hamiltonian for each spin s, the spin $-s$ appearing only in the term containing $\hat{n}_{\rho s'}$, which becomes

$$\sum_{\rho} [(\mu\mu|\rho\rho) - (\mu\rho|\rho\mu)] < \hat{n}_{\rho,s} > + \sum_{\rho} (\mu\mu|\rho\rho) < \hat{n}_{\rho,-s} >. \tag{4.6.20}$$

The 'populations' $<\hat{n}_{\rho,s}>$ and $<\hat{n}_{\rho,-s}>$ are determined iteratively for a given set of occupation numbers 1 and 0 of the SCF molecular orbitals with spin up and down (Cf. Berthier 1954, Pople and Nesbet 1954).

(b) As in the original Hoffmann EHT, the invariance of results with respect to a rotation of the atomic orbitals is not ensured. However, the changes are practically negligible as long as reasonable reference systems are considered; an example of an unreasonable choice being one whose axes do not correspond to the symmetry axes of a symmetric molecule. This is in contrast with the ZDO methods, where rotational invariance is ensured for any reference system by neglect of overlap and suppression of the angular parts of the basis orbitals.

Table 4.1. Coulomb Integrals J_{kl} and Exchange Integrals K_{kl} in eV.

Atom	J_{ss}	J_{sp}	J_{pp}	$J_{pp'}$	K_{sp}	$K_{pp'}$
O	15.11	15.14	15.87	13.77	3.66	1.05
N	13.97	13.65	13.71	12.05	3.05	0.83
C	11.73	11.48	11.51	10.22	2.59	0.64
H	12.85					

Table 4.2. Ionization Potentials and Electronegativities in eV[a] used in the BMV Method.

Atom	Valence state	Orbital	I[b]	$\frac{I+A}{2}$
O	$V_2s^2x^2yz$	s	32.36	
		x	15.07	
		y		9.83
N	V_3sx^2yz	s		20.75
		x	12.43	
		z		8.06
C	V_4tetetete	te	14.57	8.18
C	V_3trtrtrx	tr		8.95
		x		5.92
H	V_1s	s	13.59	7.17

[b] These values are slightly different from those of tables 1.2 and 1.3, due to different sources of the experimental values.

The parameters W_{μ}^{M} are computed from (cf. sec. 4.4.1):

$$H_{\mu\mu}^{0} = W_{\mu}^{M} + \sum_{\nu \in M} n_{\nu}^{(M)} (I_{\mu\nu} - \frac{1}{2} K_{\mu\nu}),$$

with

$$H_{\mu\mu}^{0} = -I, \qquad \text{if} \quad n_{\mu}^{(M)} = 2,$$

$$H_{\mu\mu}^{0} = -\frac{1}{2}(I + A), \quad \text{if} \quad n_{\mu}^{(M)} = 1.$$

Table 4.3. σ and π Charges in Aminoxyl and Iminoxyl Fragments.

	C*		N		O	
	δ_σ	δ_π	δ_σ	δ_π	δ_σ	δ_π
Aminoxyl	+0.15		-0.24	+0.27	-0.06	-0.27
Iminoxyl	+0.23	-0.21	+0.23	-0.08	-0.46	+0.29

* In the aminoxyl group, $\begin{matrix}C\\C\end{matrix}$> N - 0, C denotes one of the carbon directly linked to nitrogen. In the iminoxyl group, C = N - 0, C denotes the double-bonded carbon.

Table 4.4. Charge, Spin Density, and Observables for the Ion $CuCl_4^{2-}$.

	Cu	CI
Net charge RHF	+ 1.01	
Spin density UHF	- 0.1387	
Hyperfine coupling	calc. - 151.0 G	
constant (ESR)	obs. - 151.2 G	
Quadrupole coupling		calc. + 56.17 Mc/sec
constant (NQR)		obs. (30 to 50)*
Tensor g	perpendicular	parallel
	calc. 2.36	calc. 2.07
	obs. 2.2 - 2.3	obs. 2.00

* Experimental values of the coupling constants of chlorine in the complexes of transition metals of the third family.

Table 4.5 Applications of the SCF-EHT (BMV) Method

Molecules	References
Organic compounds	
Pyridine, quinoline, acridine	e
Imidazole, benzimidazole	g
Thiazole, benzothiazole	g
Cyclopropenyl cation	i
Bicyclo-1,1,1- pentane	l
Free Radicals	
Aminoxyl and iminoxyl fragments	d
Vinyl, cyclopropyl	h
Aminoxyl, ethyl and dimethyl-nitroxides	j
Bicyclo-1,1,1-pentyl	l
Metal complexes	
Ferro and ferricyanide anions	a
Cobalticyanide and cobaltitrioxalate anions	a
Nickelocyanide and nickelicyanide anions	a
Tetra- and hexacoordinated iron-porphyrin	a,c
Tetra- and hexacoordinated cobalt-corrin	b
Hydrated titanium cation	f
Bis- (2-methylallyl) metal complexes (metal = Cr, Fe, Co, Ni,)	k
Tetrachloride metal anions (metal = Cu, Mn)	m,n

a: Millié and Veillard 1965
b: Veillard and Pullman 1965 a
c: Veillard and Pullman 1965 b
d: Berthier *et al.* 1965
e: Veillard and Berthier 1966
f: De Brouckère 1967
g: Gélus *et al.* 1967
h: Ellinger *et al.* 1968
i: Berthier *et al.* 1968
j: Douady *et al.* 1969
k: De Brouckère 1970
l: Ellinger *et al.* 1971
m: Trappeniers *et al.* 1971
n: De Brouckère *et al.* 1973

4.7. Beyond One-Electron Schemes. Correlation, the PCILO Method, Excited States.

4.7.1 Model Hamiltonians and configuration interaction. - It has been already mentioned that model Hamiltonians can be used directly on a many-electron basis obtained by constructing 'configurations', *i.e.* Slater determinants or spin-adapted combinations thereof. The reason for the current emphasis on SCF calculations is thus the result of circumstances and of a few objective difficulties connected with use of a basis of configurations (configuration interaction, CI). The difficulties in question are the selection of the orbital basis on which the configurations should be constructed and the selection of the configurations to be actually taken into account. In fact, the total number of n-electron configurations for a basis of m orbitals is $\binom{2m}{n}$, the number of configurations where n_1 electrons have spin $\uparrow$, $n-n_1$ electrons have spin $\downarrow$ is $\binom{m}{n_1}\binom{m}{n-n_1}$; the number can be further reduced by constructing eigenfunctions of the total spin (Matsen 1964), but remains enormous for most atomic orbital bases. Therefore, it is necessary to consider only a few of the possible configurations; and, given the same criteria of choice, the quality of the results may depend on the basis orbitals. The reason for this is very simple. Given m orbitals, $|\mu_1\rangle, |\mu_2\rangle, |\mu_3\rangle, \ldots$, suppose one uses the first m_1 of them to build a basis of Slater determinants with electrons having opposite spin-components. If we now replace one or more of the molecular orbitals by new orbitals $|j_1\rangle, |j_2\rangle, |j_3\rangle \ldots\ldots$ which are linear combinations of the former ones, then every Slater determinant built on one or more of the new orbitals can be expressed as a linear combination of *all* the Slater determinants obtained from the old orbital basis; so that, in general, no subset of the new configuration basis will be equivalent to the set of configurations obtained from the first m_1 elements of the primitive orbital basis.

In view of the above considerations, configuration interaction techniques can be classified into two categories, those which start from the SCF approximation and those which bypass the SCF level. Moreover, they can be intended just to improve ground-state wave functions or to complete excited-state wave functions.

Whatever the type of CI calculations, the selection of the Hamiltonian does affect the above classification. It is important to keep in mind that the abbreviations SCF or HF designate methods derived from any approximate form of the many-particle Hamiltonian (cf. sec. 4.1).

There are different reasons why configuration interaction should be included in an ideal semiempirical scheme. As regards ground states, use of it is either an attractive alternative to the iterations of the SCF scheme and/or a way to include electron correlation into the computations. There are both qualitative and quantitative aspects of correlation (Herigonte 1972) and the qualitative aspects may be investigated and taken into account even in the framework of semiempirical methods. This was done for instance by Zanazzi and Torrini (1976) in connection with the Hubbard Hamiltonian, and was most extensively treated by Čižek *et al.* (1974, 1975), whose work constitutes a milestone in the history of the correlation problem. Their central idea was to generate a correlated ground-state wavefunction from a single-determinant one by a unitary operator, whose series expansion could be studied in terms of the diagrams of field theory; around that idea they made a number of studies mostly using the PPP Hamiltonian of sec. 4.5.1 as a model Hamiltonian. For instance, they analyzed the uncorrelated and the totally correlated limit of the PPP approximation (when the core potential tends to infinity or to zero, respectively). The use of Padé approximants was suggested to interpolate between the two limits (Čižek *et al.* 1975). This kind of study is of the greatest importance in revealing the qualitative effects of correlation corrections for any given situation.

In the case of a single determinant not constructed on SCF orbitals, what is formally still a treatment of the correlation problem becomes an analysis of configuration interaction. In this sense, the diagrammatic techniques of Cizék (1978; cf. also Paldus *et al.* 1977; Adams *et al.* 1977) represent a very efficient tool for the evaluation of the weights of the various configurations and hence for reducing the size of the CI matrix; they can also be used as a very effective form of a perturbation expansion, and in this sense they could provide a generalized version of the PCILO method.

The reasons why configuration interaction is so important for excited states, for instance triplets, is that it is not easy to define an SCF scheme for excited states, due to the open-shell problem. This is in fact possible by adopting appropriate average occupation numbers for the SCF orbitals to be determined (Roothaan 1960) and possibly by allowing for different orbitals for different spins (de Heer 1963); but the most straightforward approach is configuration interaction (possibly with optimization of the corresponding orbitals)

The PCILO method is one of the methods (if not the only one) which bypass the SCF stage by using a perturbational configuration interaction scheme in its place. We shall briefly describe it before considering excited-state calculations and the problems still open in that connection.

4.7.2 The semiempirical PCILO method. - The idea of treating a molecule as a system of localized bonds was the starting point of the original theory of resonance as well as of the semiempirical Del Re method. A complete method based on the same idea, but susceptible of any degree of accuracy was elaborated in several papers by Diner et al. (1969, a, b), Malrieu et al. (1969) under the abbreviation PCILO. That method is an example of methods which use directly configuration interaction both in semiempirical and ab-initio versions.

The one-particle basis of the PCILO method consists of bonding and antibonding two-centre orbitals $|j\rangle$; orbitals belonging to different bonds are not orthogonal to one another.

The many-particle basis is one of Slater determinants $|J\rangle$ obtained by occupying in various ways the given bond orbitals: the ground state is denoted by $|1\rangle$. The Hamiltonian is, of course

$$\hat{H}_{\text{PCILO}} = R'_0 + \sum_{j,k,s} H^{\text{core}}_{jk}\, \hat{a}^+_{js}\, \hat{a}_{ks} + \frac{1}{2} \sum (jk|lm)\, \hat{a}^+_{js}\, \hat{a}^+_{ks'}\, \hat{a}^+_{ms'}\, \hat{a}_{ks} \qquad (4.7.1)$$

the only specific detail being that the basis consists of non-orthogonal bond orbitals. Once the $|j\rangle$ orbitals are given, it takes just a few simple transformations to express the parameters H^{core}_{jk} and $(jk|lm)$ in terms of atomic Slater or Gaussian orbitals, so as to allow evaluation of the matrix elements of $\hat{H}$ over the many-electron basis. This is where the alternative semiempirical vs. *ab initio* comes into play. Once that difficulty has been overcome, the PCILO method uses straightforward perturbation theory to evaluate the many-electron ground state and its energy: the CI matrix is diagonalized by a perturbation expansion taking its diagonal as a zero-order step. Therefore the PCILO method is not an iterative one.

The parameter problem is solved in two steps. First, the bond orbitals are obtained from a standard Slater-orbital basis by minimizing the bond-bond overlap through construction of hybrids (Del Re 1963; cf. sec. 5.1.3). Next, the various integrals over the resulting hybrid AO's are determined either by approximation formulae or ab initio. In particular, it is possible to adopt the so-called ZDO approximation (sec.'s 4.5.1 and 5.7.1), and then a number of very elegant simplifications become possible. More generally, the approximations of the CNDO family can be introduced: we mention as a very recent example, the case of the INDO approximations (Douady *et al.*, 1976; 1978).

4.7.3. Applications of the PCILO-INDO method. - The PCILO method is especially interesting, of course, in the case of highly localized systems, where the bond-orbital description is quite satisfactory (of course, a generalization to problems like chemisorption is possible if the crystal orbitals are treated as if they were bond orbitals).

Therefore, PCILO is particularly effective with problems like inversion and rotation barriers. Very recent results (Douady *et al.* 1979), obtained with an improved choice of the lone-pair parameters of the method for hybrids (Del Re *et al.*, 1966), are listed in table 4.6

Table 4.6. PCILO-INDO results

a) Inversion barriers:

	ΔE_{calc}	ΔE_{exp}	
NH_3	5.2	5.2	
H_3O^+	0.8	1.7	
CH_3^-	18.4	16-18	in Kcal/mol
CH_3NH_2	7.0	6.	
Aziridine	20.5	(18.3)*	

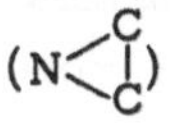

* ab initio

b) Rotation barriers:

	ΔE_{calc}	ΔE_{exp}	
$CH_3 \nrightarrow CH_3$	2.07	2.88	
$CH_3 \nrightarrow NH_2$	1.64	1.97	
$CH_3 \nrightarrow OH$	0.82	1.07	in Kcal/mole
1,3-Butadiene cis/trans	2.01	2.30	
Acetylcholine gauche/trans	3.2	3.2**	

** ab initio from Port and Pullman 1973

to give the reader an idea of the results of PCILO calculations. Those results are sometimes even qualitatively in contradiction with the standard INDO method. Table 4.7 gives an idea of computing times.

Table 4.7 Computing Times on an IBM 360/91 Computer

	PCILO INDO	SCF INDO*
NH_3	0.66 sec	0.51 sec
$H(CH_3)_3$	4.54	4.60
Acetylcholine	17.50	37.00

* program CNINDO of QCPE

4.7.4 Calculations on excited states. - In the preceding sections we have essentially referred to work on the ground states of molecules. In fact, both organic chemistry and solid-state problems (especially at the borderline level represented by clusters and chemisorption) are greatly connected with UV and visible spectroscopic properties, and therefore the excited states of free and chemisorbed molecules are most important for theoreticians.

Unfortunately, the difficulties to be overcome in applying quantum chemical (especially SCF) methods to excited states are very great, and progress has been slow. Much interesting work was done with the simplest methods, Hückel and PPP, notably by Platt (e.g. Klevens and Platt 1949a,b; A. Pullman 1949), by Parr (1964), and by many others, and many spectroscopic assignments were produced. Also the later semiempirical methods were extensively applied. Excellent books and reviews have been written on the subject; different but all-important aspects of the fundamental theory were emphasized in textbooks by Sandorfy (1964), Salem (1966), and Murrell (1971). Reviews were given among others by Platt and coworkers (1964), by Jaffé (1970), and by Richards (1974). Nearly all recent books and collections of reviews on quantum chemistry devote some space to spectroscopy. We shall not make any general presentation of the topic, but shall rather consider a few aspects which throw light on the very peculiar problems and difficulties that are found in the calculation of excited states.

First of all, assignment of bands is a very complicated problem not only when one is dealing with symmetric molecules, in which case group theory is of great help (sec. 5.8), but in heterocycles. That problem is particularly unpleasant because the use of limited bases implies that only a few energy levels are predicted, so that, when agreement with experiment is not perfect, it is impossible to be sure that the level under study is not one that is missing from the calculated set, rather than simply a level very poorly reproduced by the calculations. Moreover, approximations may cause a theoretical level to be close in energy to an experimental level (but far from it in oscillator strength), while the opposite holds for another level. Difficulties of this sort are particularly bad in connection with Rydberg states (Sandorfy, 1969; Salahub and Sandorfy 1971). A way to overcome them is to take certain molecules as reference models, and to base the classification on them. Examples of such an approach were given by Momicchioli and Rastelli (1967, 1970).

As regards a comparison of different approaches to excited states, especially triplets, we shall refer to a paper by Packer *et al.* (1969) where an attempt was made to achieve uniformity in the choice of parameters in treating both singlets and triplets and in applying different methods within the PPP scheme. From that paper we have extracted Table 4.8.

Table 4.8. First singlets and triplets of nucleotide bases (eV) above ground state by the PPP method (Packer et al. 1969).

	GS[a]	S	T		
			SCF-CI[b]	UHF[c]	Root.[d]
uracil	- 364.56	5.14	1.33	1.28	1.37
thymine	- 468.34	5.09	1.43	1.29	1.35
cytosine	- 353.25	4.41	1.51	1.45	1.74
guanine	- 586.83	4.11	1.78	1.71	1.90
adenine	- 455.44	5.00	2.72	2.96	2.82

GS = ground state S = singlet T = triplet

[a] SCF
[b] SCF plus CI of singly excited configurations
[c] spin-projected unrestricted-Hartree Fock (different orbitals for different spins)
[d] Roothaan's open shell SCF approach

The three methods illustrated there are easily described. The closed-shell SCF method for excited states just involves construction of the Slater determinants obtained by exciting one electron to a virtual SCF orbital coming from the ground-state SCF Hamiltonian, and combining them, if need be, to obtain spin-adapted combinations. The spin-projected unrestricted HF method (de Heer, 1963; Smeyers and Brucena 1976) - consists in solving the coupled eigenvalue equations corresponding to (4.1.18) written separately for spin up and spin down *without* assuming that $n_{j\uparrow} \equiv \langle \hat{n}_{j\uparrow} \rangle = n_{j\downarrow} \equiv \langle n_{j\downarrow} \rangle$, and using the occupation scheme for the state under study. As there are no doubly occupied orbitals, the resulting quantum state is a linear combination of the various states having the same S_z value $S_z = |\sum_j (n_{j\uparrow} - n_{j\downarrow})|$, e.g. 1 for a $\uparrow\uparrow$ state, and different total multiplicities $y = 2S_z + 1,\ 2S_z + e,\ 2S_z + 5, \ldots$ (3,5,7,.... if $S_z = 1$)

$$|\Phi_{S_z}^{(UHF)}\rangle = \sum_{\substack{y = 2(S_z + k) + 1 \\ k = 0,1,\ldots}} |\Phi_y^{(UHF}\rangle\, c_{S_z}^{(y)} \qquad (4.7.1)$$

with the c's suitable coefficients. Therefore, to obtain a state which is a good quantum state for the given molecule, it is necessary to transform it into a spin eigenfunction by projecting out the non-pertinent components by a projection operator defined as (Löwdin 1956)

$$\Theta_{S_z, S^2} = \prod_{l \neq S_2} \left[\frac{S^2 - l(l+1)}{s(s+1) - l(l+1)} \right] \qquad (4.7.2)$$

S^2 being the square-spin operator of eigenvalue $s(s+1)$.

Finally, as has been mentioned, Roothaan's (1960) open-shell SCF method is nothing but a derivation from the many-electron Hamiltonian (given in (4.5.1) for the PPP method) and applied to a multiconfigurational state.

An extremely important aspect of excited-state calculations is the determination of the vibrational characteristics of molecules. We shall devote the next section to certain aspects of this type of calculations, but for studies concerning electronic transitions the reader should consult the appropriate literature; as examples we mention the work of Momicchioli *et al.* on stilbene (1975) and of Lami and Del Re (1978) on the spectra of dimers.

4.8. A Case Study in Semi-Empirical Computations: Molecular Force Fields

4.8.1 Experimental material. - Experimental values are the outcome of the analysis of IR spectroscopic data, possibly supplemented by other spectroscopies: Raman, fluorescence, SVL (laser), etc...

Determinations of force constants. The 'experimental' determination of force constants consists in associating by a semiclassical calculation $M(M+1)/2$ force constants (diagonal f.c. F_{jj} and interaction f.c. F_{jk}) to the M fundamental frequencies λ_R of the vibrational spectrum. The matrix formulation of the method used to that end - the so-called Wilson method - amounts to the equation

$$\mathbf{GFL} = \mathbf{L}\Lambda \qquad (4.8.1)$$

where $\mathbf{G}$ contains the kinematic data of the problem (it is the inverse matrix of the kinetic energy coefficients over the normal modes; its elements are determined from atomic masses and interatomic distances) and $\mathbf{F}$ contains the dynamical data (force constants in the harmonic approximation); Λ is a diagonal matrix $\|\lambda_k\|$ connected to the absorption frequencies by $\lambda_R = 4\pi^2 c^2 \nu_k^2$ (ν_k = wave number in cm^{-1}), and **L** is the transformation matrix taking the normal coordinates Q_k into the internal ones R_j (**R** = **LQ**).

The inverse secular problem thus defined is not uniquely determined; the existing programs, such as Schachtschneider's, proceed from a reasonable set of initial dynamical parameters F_{jk}. The results are represented in terms of a variety of molecular force fields, obtained either by reducing the number of parameters or by imposing that they should satisfy certain relationships to one another (Table 4.9)

Table 4.9

Force field	Parameters F_{jk}
CFF (central force field)	Stretching and non-bonded atom interactions
SVFF (simple valence force field)	stretching, bending and torsion action
GVFF (generalized valence force field)	idem + various cross term
HBFF (Urey-Bradley force field)	equivalent to CFF + SVFF
OFFF (orbital following force field)	relationship between important interaction terms
HOFF (hybrid bond orbital force field)	

References:
UBFF: Urey and Bradley 1931
OFFF: Heath and Linnett 1948
HOFF: Mills 1961

For more details see Califano 1976, ch. 8.

The experimental determination of molecular force fields can be refined by taking advantage of relationships between the vibrational frequencies of molecules obtained one from the other by isotopic substitution (*e.g.* the Teller-Redlich product rule); isotopic substitution is assumed to change only the kinematic parameters of the Wilson method.

Infrared intensities. According to the classical theory of light, the integrated absorption intensity of an infrared band can be written, in the harmonic approximation

$$A_k = \frac{N}{3c^2} \sum_{\alpha = x,y,z} \left(\frac{\partial \mu^\alpha}{\partial Q_k}\right)_0^2 \qquad (4.8.2)$$

where $\left(\partial \mu^\alpha / \partial Q_k\right)_0$ represents the value at equilibrium of the α-th component of the derivative of the dipole moment with respect to the normal coordinate Q_k of the k-th vibrational mode; N and c are Avogadro's number and the velocity of light, respectively. As the normal corrdinates Q_k are related to the internal coordinates R_j through the known matrix **L**, Eq. (4.8.2) makes it possible to determine the derivatives $\left(\partial\mu^{\alpha}/\partial R_j\right)_0$ if the signs could somehow be obtained. This problem of the signs is crucial for the theory of IR spectra. Unfortunately, it can only be solved by trial and error with the help of indirect information (relative values of Coriolis coefficients ξ corresponding to the vibration-rotation coupling, changes in dipole moments observed in laser spectroscopy upon transition from the ground state to a vibrational excited state).

4.8.2 Theoretical approaches. - A theoretical work on IR spectra demands the quantum-mechanical determination of the total energy $E(\mathbb{R})$ of n electrons and N fixed nuclei as a function of the N-dimensional vector **R**, representing the nuclear configuration. In other words, it is necessary to compute the energy hypersurface of every given molecule in the Born-Oppenheimer approximation (total electronic energy plus nuclear repulsion energy). Assuming that $E(\mathbf{R})$ is known, the subsequent computations are made as follows.

Force constants. The theoretical conformation R_0 and the set of force constants of a molecule are obtained from the minimum of the energy hypersurface and from its curvature close to the minimum, respectively. For small displacements about the equilibrium configuration and up to third order in those displacements, one can write:

$$E(\mathbb{R}) = E_0 + \sum_j \left(\frac{\partial E}{\partial R_j}\right)_0 dR_j + \frac{1}{2!}\sum_{j,k}\left(\frac{\partial^2 E}{\partial R_j \partial R_k}\right)_0 dR_j\, dR_k$$

$$+ \frac{1}{3!}\left(\frac{\partial^3 E}{\partial R_j \partial R_k \partial R_l}\right)_0 dR_j\, dR_k\, dR_l + \dots \qquad (4.8.3)$$

where the first-order term is missing because at equilibrium $(\partial E/\partial R_j)_0 = 0$; $E_0 = E(R_0)$. The second derivatives at equilibrium appearing under the summation are the force constants F_{jk}.

Three types of approaches have been proposed for evaluating the F_{jk}'s. In pure analytical procedures second derivatives are obtained by differentiating the expression of the total energy with respect to the nuclear coordinates; knowledge of the second derivatives of numerical integrals is required. In analytical-numerical procedures such as the force method (Pulay 1969), the forces $-\partial E/\partial R_j$ acting in different directions are determined analytically, and the force constants are computed by differentiating numerically those forces near the equilibrium position. In purely numerical procedures, second derivatives are determined by finite difference formulas from a set of points on the potential-energy hypersurface.

Any one of the three methods above is technically practicable at least as long as semi-empirical methods are used for determining $E(\mathbb{R})$. In particular, the 'force method' was largely developed in the framework of CNDO calculations both as regards general theory (Pulay, loc. cit.) and as regards applications (Pulay and Török 1973). The results reported here (Table 10) have been obtained by methods of the third type.

First derivatives of the dipole moments. The function $\mu(\mathbf{R})$ which represents the electric dipole moment of a molecule can be expanded in the form

$$\mu^\alpha(\mathbf{R}) = \mu_0^\alpha + \sum_j \left(\frac{\partial\mu^\alpha}{\partial R_j}\right)_0 dR_j + \frac{1}{2!}\sum_{j,k}\left(\frac{\partial^2\mu^\alpha}{\partial R_j \partial R_k}\right)_0 dR_j dR_k + \ldots \qquad (4.8.4)$$

where the superscript α specifies one of the components of μ in the general coordinate system used for the molecule. The first derivatives $(\partial\mu^\alpha/\partial R_j)_0$ can be evaluated either analytically or numerically. If a finite difference scheme is adopted, where one has to extrapolate to zero results coming from finite variations ΔR_i, it is necessary to introduce a rotation correction so as to extract from the total change of the dipole moment a pure-vibration contribution which satisfies Eckart's relations (Ben Lakdar *et al.* 1978).

Numerical results. The following numerical results are extracted from a study by (E. Pouchan, Dargelos and Chaillet 1978), who compared CNDO/2, MINDO/2, PCILO, and ab initio LCAO-SCF (minimal basis STO-3G) computations.

A survey of the tables allows us to extract a number of general characteristics of force constant calculations by the three methods used.

The valence force constants (in particular the stretching diagonal force constants) are grossly overestimated in standard CNDO and PCILO whereas MINDO/2 gives a good evaluation due to its special parameter choice for energy.

As concerns bending force constants, they appear to be well predicted by CNDO/2 calculations (in the case of symmetric bending motions) and by PCILO (in the case of other motions). On the other hand, MINDO/2 leads to values which are in general too low, a result related to the fact that that method is not always capable of providing even qualitatively correct geometries, especially as regards angles. For instance the parameters proposed in MINDO/2 by Dewar and Haselbach (1970) for oxygen and nitrogen do not lead to correct values of the valence angles of those elements. Minimization of energy with respect to the geometry for water and ammonia leads to a linear form for the former and a planar form for the latter. This anomaly is due to overestimation of repulsion forces between non-linked hydrogen atoms associated with an excessive value adopted for the van der Waals radius of hydrogen (the bond lengths

Table 4.10 - Valence Force Constants (in mdy $Å^{-1}$)

Molecule		CNDO/2	PCILO	MINDO/2	exp
Acetylene	CC	34.60	38.35	18.29	15.95
	CH	15.03	12.14	6.33	6.39
Ethylene	CC	22.98	24.06	9,62	9.39
	CH_2 sym.	13.71	11.36	5.82	5.62
	CH_2 asym.	13.34	10.96	5.53	5.57
Methane	$CH(A_1)$	12.67	11.80	5.90	5.84
	$CH(F_2)$	12.05	11.20	5.30	5.38
Ethane	CC	16.23	16.05	6.12	4.45
	CH_3 sym.	12.07	11.27	5.46	4.90
	CH_3 asym.	11.56	10.70	5.01	4.76
Formaldehyde	CO	34.0	33.55	17.57	12.90
	CH_2 sym.	11.99	11.40	5.17	4.99
	CH_2 asym.	11.62	11.06	4.71	4.87
Ethane	CC/CH_3 sym.	0.671	0.704	0.443	-
valence/	CH_3sym/CH_3' sym.	0.013	-0.028	0.068	-0.007
valence	CH_3asym/CH_3' asym.	0.004	-0.005	0.004	-0.050
Formaldehyde	CO/CH_2 sym.	0.951	0.989	1.057	0.811 0.739

Table 4.11 - Bending Force Constants (in mdy $Å^{-1}$)

Molecule		CNDO/2	PCILO	MINDO/2	Exp
Acetylene	wag	0.669	0.679	0.267	0.758
Ethylene	bend	0.579	0.583	0.332	0.470
	rock	0.599	0.575	0.410	0.572
	wag	0.404	0.402	0.220	0.243
Methane	twist (E)	0.579	0.547	0.234	0.486
	bend (F_2)	0.636	0.603	0.277	0.458
Ethane	sym. bend	0.829	0.819	0.504	0.606
	asym. bend	0.798	0.779	0.387	0.560
	rock	0.794	0.797	0.521	0.682
	torsion	0.073	-	0.061	0.077
Formaldehyde	sym. bend	0.610	0.598	0.436	0.572
	rock	0.818	0.848	0.673	0.838
	wag	0.459	0.414	0.427	0.403
Ethane					
angle/angle	sym. bend/sym. bend'	0.042	0.034	0.033	0.033
angle/angle	asym. bend/asym. bend'	-0.007	-0.006	-0.007	-0.005
angle/angle	rock/rock'	0.151	0.038	0.012	0.139
angle/angle	asym. bend/rock	-0.068	-0.031	-0.068	0.007
angle/angle	asym. bend/rock'	0.013	0.005	0.024	0.004

obtained by MINDO/2 are in agreement with observed values except for CH, NH, OH which is predicted 0.1 Å too long on the average, and for NO, which is found to be too short as in CNDO/2. A fairly exhaustive table is presented by Flanigan *et al.* (1977).

The dipole moment derivatives obtained from CNDO and PCILO seem rather satisfactory, if the signs are correctly determined. The exaggeration of charge transfers typical of MINDO/2 disqualifies that method in this context.

Table 4.12. - Dipole Moment Derivatives (*).

Molecule	Vibration	CNDO	PCILO	Exp
Acetylene	asym. valence	0.294	0.283	1.170
	bending	0.845	1.385	1.373
Ethylene	asym. valence	-1.175	1.299	-0.738
	asym. in-plane bending	-0.331	-0.299	-0.03
	sym. valence	0.687	0.600	0.633
	sym. in-plane bending	0.084	0.075	0.279
Methane	valence	-0.630	-0.615	-0.833
Ethane	valence A_{2u}	0.831	0.939	1.057
	sym. bending A_{2u}	0.118	0.100	0.225
	valence E_u	-1.867	-2.035	-1.236
	rocking E_u	0.198	0.149	0.342
	asym. bending E_u	-0.645	-0.767	-0.257 -0.418
Formaldehyde	sym. valence CH	-0.57	-0.86	-0.91
	valence CO	3.46	1.69	3.39
	sym. bending	0.04	-0.07	-0.06

(*) - in $D\text{Å}^{-1}$ for symmetry coordinates of the valence type and in D for those of the bending type.

4.8.3 Force constants in π-electron theories. - In the preceding section only the 'all-valence electrons' methods have been considered, and the conclusion has been confirmed that, when those methods are required to give quantitative predictions, they behave essentially as interpolation schemes not especially reliable as soon as one gets out of the class of molecules for which they were designed. Therefore, there is no a priori reason why they should be more effective than a π-electron scheme, like the Hückel method, except perhaps for the fact that they allow for σ-π interaction.

In fact, the classical Hückel π-electron model can be used to evaluate force constants characteristic of unsaturated molecules, valence force constants of partially multiple bonds (like C $\overset{...}{-}$ C or C $\overset{...}{-}$ O stretching), twisting force constants around a (partially double bond, like out-of-plane motions for C = C and C - C bonds in conjugated systems.

Valence force constants. According to Coulson and Longuet-Higgins (1948) the changes in vibrational potential energy of a conjugated molecule can be assumed to have the form

$$\delta V = \delta F + \delta W$$

F being the π-bond energy, W being the total energy of the π system. Assuming that F is a harmonic function of the bond lengths x_{rs}

$$F = \sum_{r,s} \frac{1}{2}\sigma(x_{rs} - s)^2 \qquad (4.8.5)$$

σ and s being the force constant and the length of the purely single bond. W depends on bond lengths through the bond integrals β_{rs}, and, following Lennard-Jones, one writes

$$2\beta_{rs} = \frac{1}{2}\xi\left(x_{rs} - d\right)^2 - \frac{1}{2}\sigma\left(x_{rs} - s\right)^2 + C \qquad (4.8.6)$$

ξ and d being the force constant and the length of the purely double bond.

By deriving V with respect to the distance x_{rs} the internuclear equilibrium constant is found to be

$$x_{rs} = \frac{d\xi p_{rs} + s\sigma(1 - p_{rs})}{\xi p_{rs} + \sigma(1 - p_{rs})} \qquad (4.8.7)$$

where p_{rs} is a π-electron bond order given by

$$\frac{1}{2}\frac{\partial W}{\partial \beta_{rs}} = p_{rs} \qquad (4.8.8)$$

Further derivation produces the force constant

$$k_{rs} = \left\{\sigma(1 - p_{rs}) + \xi p_{rs}\right\} + \left\{\frac{\xi\sigma(s - d)}{\sigma(1 - p_{rs}) + \xi p_{rs}}\right\}^2 \frac{\pi_{rs,rs}}{2} \tag{4.8.9}$$

where $\pi_{rs,rs}$ is a π-electron polarizability given by

$$\frac{1}{2}\frac{\partial^2 W}{\partial \beta_{rs}^2} = \pi_{rs,rs} \tag{4.8.10}$$

Numerically it is found that $\pi_{rs,rs}$ is a monotonically decreasing function of p_{rs}, which goes to zero for a strictly double bond, for which $p_{rs} = 1$. Therefore, the contribution of the second term of (4.8.9) to k_{rs} is generally weak and anyway a function of p_{rs}; therefore, the theoretical expressions obtained for the valence force constants in unsaturated molecules as well as the expression for internuclear distances are formulas for interpolation between the single-bond and the double-bond limit.

Equation (4.8.9) has been used mainly to predict the vibrational frequencies of chemical groups behaving as independent vibrators. For instance, in the case of the carbonyl group, the relationship

$$\nu = \frac{1}{2\pi}\sqrt{\frac{k_{C=O}}{M}} \tag{4.8.11}$$

where M is an effective mass, then makes possible a study of the characteristic frequencies $\nu_{C=O}$ in aromatic aldehydes, ketones, and quinones (Berthier *et al.* 1952) as shown in table 4.13.

Twisting force constants. According to Anno (1958) the torsional out-of-plane motions can be studied just on the basis of the total π-energy W. It is supposed that W depends only on the bond integral β_{rs} of the bond about which the twisting takes place, and that

$$\beta_{rs} = (\beta_{rs})_0 \cos\varphi, \quad (\beta_{rs})_0 = k(s_{rs})_0, \tag{4.8.12}$$

where φ denotes the angle of torsion, $(\beta_{rs})_0$ the bond integral for the planar equilibrium position, and $(s_{rs})_0$ the corresponding overlap integral.

With the above assumptions, the twisting force constant in the r,s bond is

$$\theta_{rs} = \frac{\partial^2 W}{\partial \beta_{rs}^2}\frac{\partial^2 \beta_{rs}}{\partial \varphi^2} = -2\, p_{rs}(\beta_{rs})_0 = k\, p_{rs}(s_{rs})_0 \tag{4.8.13}$$

Table 4.13 - Hückel-Method Analysis of the IR Frequency of the Carbonyl Group in Aromatic Compounds. (*)

COMPOUNDS	$p_{C=O}$	$\nu_{C=O}$	k_{theor}	ν_{theor}	ν_{exp}
Formaldehyde	0.958	0.048	13.47	(1744)	1744
Glyoxal	0.937	0.044	13.29	1732	1730
Benzaldehyde	0.905	0.076	12.93	1708	1708
Acrolein	0.895	0.085	12.81	1700	1700
Benzophenone	0.857	0.105	12.41	1674	1664
Benzonaphthenone	0.820	0.146	11.96	1643	1644
Cyclopentadienone	0.895	0.063	12.86	1703	----
Fluorenone	0.871	0.092	12.57	1684	1718
Dibenz-1,2,7,3-fluorenone	0.869	0.097	12.54	1682	1694
Tropone	0.779	0.172	11.50	1611	1638
Benz-4,5-tropone	0.799	0.157	11.73	1627	1641
Dibenz-2,3,6,7-tropone	0.834	0.128	12.14	1655	1660
Parabenzoquinone	0.856	0.104	12.40	1673	1667
Anthraquinone-9,10	0.860	0.102	12.45	1676	1679
Dibenz-1,2,5,6 anthraquinone 9,10	0.858	≅0.11	12.42	1674	1663
Diphenoquinone	0.819	0.159	11.92	1640	1626
Pyrenequinone-3,10	0.833	0.137	12.11	1652	1645
Orthobenzoquinone	0.879	0.090	12.65	1690	----
Phenanthrenequinone-9,10	0.885	0.082	12.73	1694	1683
Acenaphthenequinone	0.887	0.081	12.75	1696	1736

(*) Symbols: p = bond order; π = polarizability; k = force constant; ν = IR frequency (in cm^{-1})

If the proportionality constant is calculated on the observed values in ethylene and berzene the values $\theta_{C-C} = 0.0951$ and $\theta_{C=C} = 0.5209$ are found for butadiene, to be compared with 0.5382 for ethylene (unit: 10^{-11} erg/rad).

4.9. Limitations of Semiempirical and Limited-basis Methods: The N_2O_4 Molecule.

There are circumstances where a very delicate balance of effects is responsible for the observed properties. In those cases only very elaborate calculations provide reasonable results. We illustrate this by a specific example.

The molecule N_2O_4 presents several peculiar features; the length of the bond N-N between the two moieties is unusually large (1.782 Å (McClelland 1972) to be compared with 1.47 Å, length of the N-N bond in the hydrazine N_2H_4 or with 1.49 Å in N_2F_4 (Gilbert et al. 1972)); the enthalpy of homolytic dissociation is small (Boilar et al. 1973) (12.9 kcal. $mole^{-1}$ instead of 60 ± 4 kcal. $mole^{-1}$ in N_2H_4). The barrier to internal rotation about N-N is higher than would be expected for such a long bond, being 2.9 kcal. $mole^{-1}$ with a torsional frequency of approximately

50 cm^{-1} (Snyder and Hisatsune 1957). The molecule N_2O_4 is diamagnetic whereas the building unit is paramagnetic.

The structure of the molecule has been extensively investigated; pure N_2O_4 in solid phase (Snyder and Hisatsune 1957; Begun and Fletcher 1960) at not-too-low temperature, in the liquid phase (Snyder and Hisatsune 1957; Begun and Fletcher 1960) and in the gas phase (Snyder and Hisatsune 1957; Smith and Hedberg 1956) consists of planar molecules (symmetry D_{2h}) having the configuration O_2N-NO_2. The existence of molecules with the same configuration but staggered (symmetry D_{2h}) and of molecules with the configuration O_2N-ONO has also been reported in N_2O_4 solid at liquid nitrogen temperatures (Hisatsune *et al.* 1960) The results for the coplanar molecule show a great similarity with those found for NO_2 itself (McLelland *et al.* 1972):

$$\begin{cases} r_{NO} = 1.190\ \text{Å in } N_2O_4,\ 1.202\ \text{Å in } NO_2 \\ < ONO = 135.4^\circ \text{ in } N_2O_4,\ 123.02^\circ \text{ in } NO_2 \end{cases}$$

It was therefore a real theoretical challenge to explain the very long, weak N-N bond and the planarity of the molecule although the O-O repulsions are a maximum for this configuration, as well as all the other peculiar properties of N_2O_4.

The N-N bonding in N_2O_4 has been variously considered: Chalvet and Daudel (1952) assumed that the nitrogen and oxygen atoms are in sp^2 hybridization; they included eight π-electrons and interpreted the great length of the N-N bond as due to the repulsion between nitrogen atoms with a partial positive charge. They obtain a total N-N bond order of 1.64 in N_2O_4; this value seems too large for the length of the N-N bond. Coulson and Duchesne (1957) gave a quite different interpretation of this abnormal bond length and suggested considering it as a π-only system with the σ and σ^* levels both occupied to give a net π-bond order of 0.3. McEwen (1960) suggested charge-transfer and dispersion type interactions between the π-electron systems of the two fragments. Valence bond and "double quartet" (Green and Linnett 1960) descriptions of the molecule have also been given. In its valence bond description Pauling (1960) has rejected the idea that there is no σ-bond and just a π-bond. For Pauling the unpaired electron in NO_2 clearly occupies a σ-MO and consequently a σ-bond should be formed. As the odd electron in NO_2 is delocalized and is located only 42 % on the nitrogen atom, one would expect a bond strength of 0.42 which according to Pauling would correspond to a bond length of about 1.75 Å. This result of 0.42 agrees reasonably well with the value of 0.34 given by the bond length. Pauling assumed that there is about 4 % conjugation of the two N = O bonds and that this amount of conjugation gives the N-N bond enough double-bond character to require the observed planarity. We shall see that this description of N_2O_4 is very close to the one obtained by Ahlrichs and Keil (1974) after the most complete ab initio computation performed to this day.

The extended Hückel calculations of Green and Linnett (1960) for NO_2 and N_2O_4 suggested that the dimer was stabilized by σ-bonding plus π-bonding both in and out of the molecular plane. A computation by the PPP method (Serre 1961) showed that the π-only model led to the prediction of a triplet ground state and was therefore in contradiction with the observed diamagnetism of the molecule.

Several calculations have been attempted by Brown and Harcourt (1961), Harcourt (1968) using the variable electronegativity self-consistent field (VESCF) method. They proposed a structure similar to the sigma plus pi structure with eight pi-electrons and with delocalization of oxygen lone-pair electrons into the antibonding N-N sigma orbital. Moore (1967) by an EHT program found a value of 2.11 kcal. $mole^{-1}$ for the barrier to internal rotation, compared with the experimental value of 2.9 kcal. $mole^{-1}$. He correctly found that the stable form is planar. Redmond and Wayland (1968), proposed on the basis of Extended Hückel Theory and CNDO/2 calculations that the barrier to rotation was the result of interactions other than N-N π-bonding, particularly long-range O-O σ-interactions and the dependence of the N-O bonding on the dihedral angle. The CNDO/2 method was also used by Kelkar et al. (1971). Their calculations correctly predicted the molecule to be diamagnetic but the calculated bond length was only 1.40 Å (experimental value: 1.75 Å) and the internal rotation barrier was only 1.4 kcal/mole (experimental value: 2.9 kcal. $mole^{-1}$). The most recent semiempirical calculation has been performed by Harcourt (1979) by the CNDO method and by Kishner *et al.* (1978) by the CNDO/BW method (Kishner *et al.* 1968). This is a version of the CNDO/2 method in which bonding parameters and core repulsions are determined by extensive optimization of bond energies and bond lengths for each atom pair. The localized molecular orbitals (LMO) derived from this CNDO/BW calculation correspond (1) to an oxygen lone pair, mainly of character 2s (2) to a N-O σ bond with 61.5 % of nitrogen orbital (3) to a second oxygen lone pair, delocalized to the p_x orbitals of the nitrogen atoms (4) to an oxygen lono pair, giving a highly polar N-O π bond (5) to an N-N σ bond with no π character. The main feature of this semiempirical calculation is that the orbital of the second oxygen lone pair is antibonding, reducing the N-N bond strength and giving a weak N-N bond order of 0.515. After such a delocalization, the charges are -0.30 on the oxygen atoms and +0.60 on the nitrogen atoms. These latter charges cause an electrostatic repulsion and explain the long and weak N-N bond. The main results of this calculation are summarized in Fig. 1. The minimization of the total energy with bond length gives a theoretical N-N bond length of 1.46 Å, compared 1.78 Å experimentally. The CNDO/BW wrongly predicts that N_2O_4 is unstable relative to $2NO_2$ but gives a reasonable binding energy (425 kcal.

mole^{-1} at 1.46 Å in good agreement with the 455 kcal. mole^{-1} from thermochemical measurements).

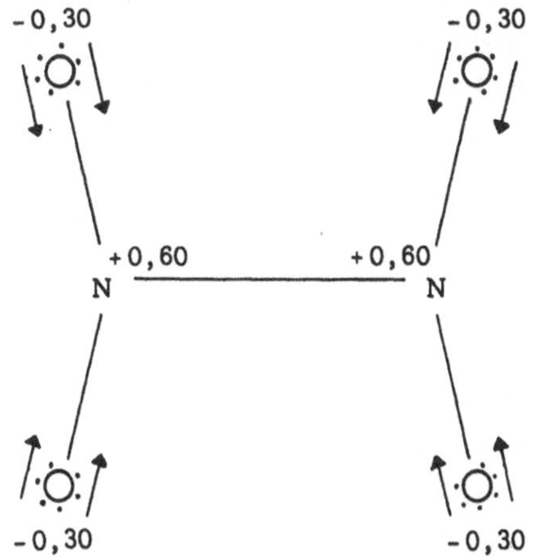

Fig. 1

Thus, the N_2O_4 problem was not solved by semiempirical methods. In recent years several "ab initio" computations appeared. From a Hartree-Fock LCAO MO calculation with a basis of symmetry orbitals formed from a minimum Slater basis set, Griffiths *et al.* (1974) found the lowest energy state to be N-N antibonding stabilized by N-N-O 3-center interactions; the N-N π bonding was not large. Even after a limited CI they found N_2O_4 unstable relative to two NO_2. In fact, this calculation disagrees in ordering the 12 highest occupied orbitals compared to the other ab initio calculations and to EHT and CNDO computations; in addition there is one level of unique symmetry. It has been shown by photoelectron spectroscopy (Ames and Turner 1976) that σ bonding must exist since there is an increase in vertical ionization energy from 11.24 eV in NO_2 to 11.6 eV in N_2O_4. The two other ab initio calculations confirmed this experimental result. By ab initio calculations using an extended Gaussian basis set, Howell and Van Wazer (1974) suggested that the planar geometry is due to O-O σ-interactions and that there is donation of the oxygen lone pairs into the N-N σ* orbital. These authors also performed Extended Hückel calculations for the ab initio optimized geometries. The results of these latter calculations were roughly similar to those of the ab initio studies. The final optimized value for the N-N distance was 1.67 Å. By shifting from the (52/52) to the (73/73) basis set the optimized N-N bond distance increases by about 0.02 Å. The rotational barrier calculated for N_2O_4 is 11.6 kcal. mole^{-1} for the larger basis set. The calculation assigns a planar structure to the N_2O_4 molecule. The alternative electronic configurations previously suggested by Coulson and Duchesne (1957) for the ground state of N_2O_4 were also explored and they were found to be dissociative.

Ahlrichs and Keil (1974) performed a series of ab initio computations for various N-N distances for the planar (D_{2h}) and the skewed (D_{2d}) structure in using an extended basis set. Since the Hartree-Fock method cannot describe accurately the dissociation of an electron pair bond into the open shell fragments, they partially included the effects of electron correlation. Since only one electron pair bond is broken in the dissociation $N_2O_4 \rightarrow 2\ NO_2$, they only account for the electron correlation of the valence electron pair. This can be done by replacing the MO describing the bond N-N by a two-electron configuration - interaction (CI) function describing the valence pair and by using an expression of this function in natural orbitals. The basis set was of "double" quality and a *d* set was included for the nitrogen atom. The basis set employed reproduced the experimental structure of NO_2 up to 0.01 Å and ± 1°. In a first series of computations (s and p functions only), a N-N distance of 1.67 Å was found on the HF level and of 1.81 Å if the valence pair correlation was included. A comparison of the energy of N_2O_4 at the end of this calculation with the RHF energy of NO_2 showed that N_2O_4 was not predicted to be stable with respect to 2 NO_2 for this basis set. With the d set on nitrogen, for the planar geometry, the minimum of the potential curve is at a N-N distance of 1.59 Å on the HF level and at 1.67 Å if the valence pair correlation is included. The molecule N_2O_4 is still predicted to be unstable with respect to 2 NO_2. Including the valence pair correlation they at last get a binding energy of 5.0 kcal. mole^{-1} (experimental value: 12.7 kcal. mole^{-1}). For the skew configuration, with the d set on nitrogen, they get a minimum of energy with correlation for 1.62 Å and this energy is still 1 kcal. mole^{-1} higher than for 2 NO_2.

With such an introduction of the correlation energy it seems difficult to obtain a better agreement between the experimental barrier (2.9 kcal. mole^{-1}) and the computed one (6.0 kcal. mole^{-1} in this calculation with the d set on nitrogen). In fact, Ahlrichs and Keil (1974) found that the inclusion of electron correlation has little influence on electron distribution in N_2O_4. The inclusion of the valence pair correlation results essentially in an admixture of the antibonding b_{1u} MO; this molecular orbital approximates well to the antibonding linear combination of the singly occupied molecular orbitals of the NO_2 fragments. The configuration in which this orbital is occupied by two electrons plays an important role in the computation of the binding energy but it contributes little to the wave function since the corresponding occupation number is only 0.02.

In N_2O_4, the weakness of the N-N bond can have two reasons; the first one is a relatively small N-N bond order of the highest occupied MO which essentially describes the bond between the two NO_2 moieties; the second one is the repulsion of the doubly occupied molecular orbitals of NO_2.

The form of the NO_2 orbital corresponding to the highest occupied MO in N_2O_4 proves that the delocalization of the unpaired electron of NO_2 extends over the whole molecule. No significant change takes place on the formation of N_2O_4 from 2 NO_2. But this σ-type MO is not spherically symmetric with respect to the N-N axis and the bonding is stronger in the planar geometry of N_2O_4.

The weakness of the N-N bond is due to the repulsion of the doubly occupied MO's of NO_2, to the delocalization of the bond electron pair over the whole molecule and to a shift of charge from N to O, which weakens the N-N bond.

The coplanarity of N_2O_4 results from a delicate balance between two effects, the non-spherical character of the $6a_g$ MO and the weak O-O bond favour the planar structure but the repulsion of the doubly occupied MO's of the NO_2 fragments favours the skew configuration.

We see that the peculiar features of N_2O_4 are now explained; but quantum chemists had to wait for calculations with large basis sets and with introduction of correlation energy before really understanding them. This is in striking contrast with cases, like that of azulene (sec. 5.4.3), where sophisticated computations have contributed changes in numerical values of no significance for the interpretation of facts.

Chapter 5. The Basis Problem

5.1 The MVAO Basis

The simplest semi-empirical models hinge on the choice of the MVAO basis. We have already mentioned that, in the framework of those models, each first-row atom contributes four atomic orbitals to the basis, hydrogen contributes one orbital, etc.; the resulting set is a "Modified Valence Atomic Orbital" (MVAO) basis. This corresponds to the idea that every first-row atom can lose or take up to eight electrons, either by forming an ion or by filling orbitals which result from combining each of the four orbitals with one or more orbitals of other atoms.

The formation of a bond is thus described by mixing of one orbital per atom and filling of the lower bond orbital (or of one of the lower delocalized orbitals) by two electrons. This simple recipe implies a number of assumptions which combine with the above general limitation to restrict the choice of the atomic orbitals to be used.

5.1.1 Localization: A specific example. - Before proceeding to general considerations, we examine again a simple method for saturated molecules. The whole procedure is based on the one-electron effective Hamiltonian which, in our model, determines the form of the molecular orbitals. If a block can be factorized out of the Hamiltonian matrix in question, there will be a certain number of orbitals which are formed with a subset of the given basis, and are formally completely independent of the rest of the molecular orbitals. For instance, in a molecule like CO_2, the MVAO basis has the form

$$|\mu\rangle \equiv (|1\rangle\, |2\rangle \ldots |\dot{\dot{1}}\dot{\dot{2}}\rangle \ldots |12\rangle) - (|l_1\rangle\, |l_2\rangle\, |h_0\rangle\, |p\pi_0\rangle$$
$$|h_c\rangle\, |h_{c'}\rangle\, |p\pi_c\rangle\, |p\pi_c'\rangle\, |p\pi_0'\rangle\, |h_0'\rangle$$
$$|l_1'\rangle\, |l_2'\rangle) \qquad (5.1.1)$$

where the prime refers both to the right hand oxygen and to the plane perpendicular to that of the π orbitals; the letter l stands for lone pairs, the letter h for the hybrids.

Suppose now that the effective Hamiltonian can be represented by a matrix $\mathbf{H}$ which consists of 1×1 and 2×2 blocks corresponding respectively to the lone pairs and to the four σ and π bonds of the molecule.

Then the whole system can be treated as if it consisted of eight independent electron pairs, to be described in terms of smaller bases of one or two elements (Equation (5.1.5)).

With the above asssumption, the eigenvalues for the system are those of the matrix

$$\mathbf{H} = \begin{vmatrix} H_{1,1} & 0 & 0 & 0 & 0 & 0 & 0 & 0 & 0 & 0 & 0 & 0 \\ 0 & H_{2,2} & 0 & 0 & 0 & 0 & 0 & 0 & 0 & 0 & 0 & 0 \\ 0 & 0 & H_{3,3} & 0 & H_{3,5} & 0 & 0 & 0 & 0 & 0 & 0 & 0 \\ 0 & 0 & 0 & H_{4,4} & 0 & 0 & H_{4,7} & 0 & 0 & 0 & 0 & 0 \\ 0 & 0 & H_{5,3} & 0 & H_{5,5} & 0 & 0 & 0 & 0 & 0 & 0 & 0 \\ 0 & 0 & 0 & 0 & 0 & H_{6,6} & 0 & 0 & 0 & H_{6,10} & 0 & 0 \\ 0 & 0 & 0 & H_{7,4} & 0 & 0 & H_{7,7} & 0 & 0 & 0 & 0 & 0 \\ 0 & 0 & 0 & 0 & 0 & 0 & 0 & H_{8,8} & H_{8,9} & 0 & 0 & 0 \\ 0 & 0 & 0 & 0 & 0 & 0 & 0 & H_{9,8} & H_{9,9} & 0 & 0 & 0 \\ 0 & 0 & 0 & 0 & 0 & H_{10,6} & 0 & 0 & 0 & H_{10,10} & 0 & 0 \\ 0 & 0 & 0 & 0 & 0 & 0 & 0 & 0 & 0 & 0 & H_{11,11} & 0 \\ 0 & 0 & 0 & 0 & 0 & 0 & 0 & 0 & 0 & 0 & 0 & H_{12,12} \end{vmatrix} \qquad (5.1.2)$$

where, in addition to four 1×1 blocks, there are four fully independent 2×2 blocks

$$\mathbf{H}(\sigma) = \begin{vmatrix} H_{3,3} & H_{3,5} \\ H_{5,3} & H_{5,5} \end{vmatrix} \text{ and } \mathbf{H}(\sigma') = \begin{vmatrix} H_{6,6} & H_{6,10} \\ H_{10,6} & H_{10,10} \end{vmatrix} \text{ for the two } \sigma \text{ bonds;} \qquad (5.1.3)$$

$$\mathbf{H}(\pi) = \begin{vmatrix} H_{4,4} & H_{4,7} \\ H_{7,4} & H_{7,7} \end{vmatrix} \text{ and } \mathbf{H}(\pi') = \begin{vmatrix} H_{8,8} & H_{8,9} \\ H_{9,8} & H_{9,9} \end{vmatrix} \text{ for the two } \pi \text{ bonds,} \qquad (5.1.4)$$

so that the lone pair molecular orbitals are just the corresponding basis elements, and the bonding orbitals have the general form

$$|j\rangle = c_{\mu j}|\mu\rangle + c_{\nu j}|\nu\rangle,$$
$$(|\mu\rangle, |\nu\rangle \ \sigma \text{ or } \pi \text{ orbital of linked atoms}). \qquad (5.1.5)$$

We shall say that we have localized our system when we have factorized our effective Hamiltonian into a form of the type (5.1.3) or (5.1.4), possibly with blocks of order higher than 2, as will happen, for instance, in the case of benzene. Localization is introduced at the (model or effective) Hamiltonian level, not at the wave function level.

The factorization under study is not imposed in principle, but is expected to result from the geometry of the given molecule and from the properties of the basis; which, for the purpose, can be built from spherical-harmonic orbitals by suitable hybridization, as will be discussed later, but of course should not be artificially localized.

Except in cases of different symmetries, the factorization will be only a first order property. Moreover, even if factorization held perfectly, it would not imply that the various subsystems into which the molecule is divided are actually independent of one another. The Hamiltonian *operator* which enters each of the matrix elements depends anyway on the whole molecule.

5.1.2 Localization and Overlap. - It is possible to see that a form of the type (5.1.5) implies some kind of restriction on the atomic orbitals of the basis by considering the non-orthogonality problem.

Let us suppose that the effective bond Hamiltonian can be written in configuration space, in the form

$$\hat{H}_1^{\mathrm{eff}} = \hat{\mathcal{T}} + \sum_Z \hat{V}^{\mathrm{eff}}(Z) = \frac{1}{2}\left\{\hat{H}_1^{\mathrm{eff}}(X) + \hat{H}_2^{\mathrm{eff}}(Y)\right\} + \sum_{Z \neq X, Y} \hat{V}^{\mathrm{eff}}(Z) \qquad (5.1.6)$$

where

$$\hat{H}_1^{\mathrm{eff}}(X) = \hat{\mathcal{T}} + 2\hat{V}^{\mathrm{eff}}(X). \qquad (5.1.7)$$

We require for $|X(Y)>$ and $|Y(W)>$:

$$<X(Y)|\hat{H}_1^{\mathrm{eff}}|X(Y)> = \varepsilon_{X(Y)}^{\mathrm{eff}} + <X(Y)|\sum_Z (\hat{V}^{\mathrm{eff}}(Z) - \hat{V}^{\mathrm{eff}}(X))|X(Y)> \qquad (5.1.8a)$$

$$<Y(W)|\hat{H}_1^{\mathrm{eff}}|X(Y)> = \frac{1}{2}\left(\varepsilon_{X(Y)}^{\mathrm{eff}} + \varepsilon_{Y(W)}^{\mathrm{eff}}\right)<Y(W)|X(Y)> + \Delta_{X(Y),Y(W)} \qquad (5.1.8b)$$

if (a) the orbital $|X(Y)>$ satisfies the equation

$$\hat{H}_1^{\mathrm{eff}}(X)|X(Y)> = \varepsilon_{X(Y)}^{\mathrm{eff}}|X(Y)>, \qquad (5.1.9)$$

(b) the (four) orbitals of X are degenerate:

$$\varepsilon_{X(Y)}^{\mathrm{eff}} = \varepsilon_X^{\mathrm{eff}}, \qquad (5.1.10)$$

and (c) $\Delta_{X(Y),Y(W)}$ is negligible.

Eq. (5.1.8b) can then be written

$$< Y(W)|\hat{H}_1^{eff}|X(Y)> = \lambda_{XY}< Y(W)|X(Y)>. \qquad (5.1.11)$$

Evidently, the block factorization of $\hat{H}_1^{eff}$ shown in Eq. (5.1.2) is possible only if

$$< Y(W)|X(Y)> = 0, \text{ unless } W = X. \qquad (5.1.12)$$

The two equations (5.1.9) and (5.1.12) are requirements on the basis atomic orbitals, and ensure in particular that (5.1.11) be satisfied. Consider in particular requirement (5.1.12). This is certainly satisfied if the atomic orbitals form an orthogonal set; but, if so, the bond integrals given in (5.1.11) all vanish; and, at any rate, our atomic orbitals are defined up to a spherical harmonic by Eq.s (5.1.9) and (5.1.10). Those equations provide the only degrees of freedom available; it is possible to take advantage of them to satisfy as closely as possible Eq. (5.1.12). This leads to the construction of hybrids according to the maximum localization criterion (Del Re 1963c).

5.1.3 Hybridization and localization. - It is well known that localization can be realized in an infinite number of ways; this is what led some to claim that the bond concept is an arbitrary one. One can eliminate the arbitrariness by introducing hybridization according to the MO version of Pauling's idea: Combine the pure MVAO's for each atom separately so as to obtain, if possible, orthogonal *hybrid* orbitals having as small an overlap as possible with all the other hybrid orbitals except one; The pairs thus obtained will define the bonds. If no matter how hybridization is realized more than two hybrids have overlaps of the same order of magnitude, we shall say that we have a many-centre bond.

The procedure to realize the scheme outlined above is as follows:
(a) Compute the overlap matrix **S** for the pure basis defined by (5.1.9):

$$S_{\mu\nu} \equiv < \mu|\nu> \qquad (5.1.13)$$

and arrange it in blocks $\mathbf{S}^{(XY)}$ associated with pairs of atoms; (b) diagonalize by means of unitary matrices $\mathbf{D}^{(XY)}$ each of the $\mathbf{S}^{(XY)}$ blocks via the symmetric matrix $\mathbf{S}^{+}\mathbf{S}$:

$$\mathbf{D}^{(XY)+}\mathbf{S}^{(XY)+}\mathbf{S}^{(XY)}\mathbf{D}^{(XY)} = \text{diag}. \qquad (5.1.14)$$

$$\mathbf{D}^{(YX)+}\mathbf{S}^{(XY)}\mathbf{S}^{(XY)+}\mathbf{D}^{(YX)} = \text{diag}. \qquad (5.1.15)$$

(c) for each atom X of valence v select the v highest eigenvalues of Eq. (5.1.14) allowing Y to be any atom in the molecule (this is the step which decides where the bonds are) and call $\mathbf{D}_{\xi(Y)}$ the columns of the $\mathbf{D}^{(XY)}$ matrices (X fixed) which correspond to the highest eigenvalues denoted by $\lambda^2_{\xi(Y)}$; (d) construct v *orthogonal* columns $\mathbf{U}^{(X)}_{\xi(Y)}$ using the condition that for equal λ^2 they should be as close as possible to the corresponding $\mathbf{D}_{\xi(Y)}$'s, and the closer to $\mathbf{D}_{\xi(Y)}$ the higher the corresponding λ^2 value; and (e) build a block-diagonal unitary transformation U where each block $\mathbf{U}^{(X)}$ is obtained from the set of v columns $\mathbf{U}^{(X)}_{\xi(Y)}$ just obtained.

If the orthogonality constraints were not there, the above procedure would give a unitary transformation of the pure MVAO basis which mixes with one another only the MVAO's of the same atom (hybridization) and makes most of the elements of the new overlap matrix as small as is possible with this kind of transformation. Orthogonalization reduces the quality of the results, but usually gives good results: In ordinary saturated molecules it does happen that overlaps corresponding to bonds of the chemical formula are large.

(One exception is given by hydrogen-hydrogen overlap in methyl and methylene groups, and this may indicate either that the Slater AO's used are not good approximations to correct MVAO's or that CH bonds should not be treated separately.)

The procedure outlined above has the quality of not requiring knowledge of where bonds are; indeed, an orbital-following procedure can be based on it so as to predict geometries (Pozzoli *et al.* 1973).

5.2 The Non-orthogonality Problem

The above considerations suggest that the assumption that the MVAO's are orthogonal to one another, as made in the original methods of the Hückel type and in more recent semiempirical methods, is not compatible with the very *atomic* nature of the basis orbitals. This difficulty was overcome long ago by Löwdin, who showed that it is possible to construct an orthogonalized set of quasi-atomic orbitals; by admixing small contributions of atomic orbitals of other centers to the atomic orbital of a given centre it is possible to construct a set of orthogonal orbitals which differ surprisingly little from the original atomic orbitals.

We shall not stop on this problem, but recall the Löwdin transformation (Löwdin 1950): the LOAO (Löwdin orthogonalized atomic orbital) basis $|\underline{\chi}\rangle$ given by the transformation

$$|\underline{\chi}\rangle = |\underline{\mu}\rangle\, \mathbf{S}^{-\frac{1}{2}} \qquad (5.2.1)$$

$|\mu>$ being the original basis and $\mathbf{S}$ its overlap matrix. Of course, the Löwdin transformation may give wildly delocalized orbitals if the elements of $\mathbf{S}$ are very large. However, strongly overlapping orbitals, which mix with one another, would do so anyway in the molecular orbitals, in as much as they belong to non-localizable subsystems of the given molecule; and, especially in case of symmetric systems, they can still be interpreted as being associated with individual atoms. Nevertheless, it is advisable to analyze and reformulate all semiempirical methods in terms of overlapping orbitals especially when more than two centers are involved in the molecular orbitals; unless one is confronted with the non-orthogonality catastrophe mentioned below.

5.2.1 The Hückel method in a non-orthogonal basis. - A very simple example of qualitative differences obtained when overlap is included is provided by the Hückel method when bond integrals are assumed to be proportional to the corresponding overlap integrals; this circumstance corresponds to precise assumptions on Eq. (5.1.8). The general case has been treated by Del Re (1960). Here let us assume that $\mathbf{H}$ and $\mathbf{S}$ commute, and hence have the same eigenvector matrix $\mathbf{T}$, with diagonal forms $\mathbf{h}$ and $\mathbf{s}$, respectively.

Suppose, for instance, that

$$\mathbf{s} = \mathbf{I} + \lambda\,\mathbf{h} \tag{5.2.2}$$

($\mathbf{I}$ being the unit matrix) with an $\mathbf{H}$ whose diagonal elements are all zero. Then

$$\mathbf{H\,C} = \mathbf{SC}\varepsilon \tag{5.2.3}$$

becomes

$$\mathbf{S}^{-1}\mathbf{HC} = \mathbf{C}\,\varepsilon \tag{5.2.4}$$

This equation is satisfied by

$$\varepsilon = \mathbf{s}^{-1}\mathbf{h} \text{ i.e. } \varepsilon_{jj} = \frac{h_{jj}}{1 + \lambda h_{jj}} \tag{5.2.5}$$

Thus ε is the eigenvalue matrix of $\mathbf{S}^{-1}\,\mathbf{H}$; and, unless $\mathbf{S} = \mathbf{1}$, one obtains eigenvalues which do not share certain properties of the eigenvalues of $\mathbf{H}$. For instance, if the eigenvalues of $\mathbf{H}$ are symmetric with respect to zero, the ε_{jj}'s are not; which is very important when electron affinities are discussed.

The choice (5.2.2) is valid for the Hückel method as applied to benzene when the *zero-point* is the atomic integral of carbon. Then the Hückel method with overlap (S = 0.25) gives the eigenvalues shown in table 5.1, second line.

Table 5.1

Eigenvalues of Benzene

	X_1	X_2	X_3	X_4
$\lambda = 0$	2	1 , 1	-1,-1	-2
$\lambda = 0.25$	1.333	0.800	-1.3333	-4.000
$\lambda = 0.65$	0.8696	0.6061	-2.8571	6.6667

5.2.2 The non-orthogonality catastrophe and the nearest-neighbor approximation:
A dramatic conclusion that can be extracted from Eq. (5.2.6) is that strictly speaking non-orthogonality of the basis and the nearest-neighbor approximation are not compatible. In fact, the denominator of ε_{jj} can be zero if the **H** matrix possesses an eigenvalue h_{jj} equal to $-1/\lambda$. If there are eigenvalues on either side of that value, the ε's are no longer in a simple correspondence with the h's; which is not an acceptable situation. An example is provided by a benzene-like system with $\lambda = .65$ (Table 5.1, last line): The off-diagonal elements of the overlap matrix will be still lower than 1 - and thus physically acceptable - but the correspondence of table 5.1 becomes very peculiar. Yet, in the case of metal crystals overlaps larger than 0.5 between nearest neighbors are quite reasonable.

The origin of the surprising result just mentioned is really quite simple; equation (5.2.2) is theoretically unacceptable when λ is such that one or more of the elements of **s** are negative, because by definition the overlap matrix is a positive definite matrix (Löwdin 1956). Now, the two approximations consisting in a proportionality rule for hopping integrals and in neglect of non-nearest neighbor hopping integrals lead in many cases - just as in the example of Table 5.1 - to assuming negative eigenvalues for the overlap matrix. An important conclusion is that the nearest-neighbor approximation is generally incompatible with the assumption that to AO basis is a non-orthogonal one (Del Re *et al.* 1977a).

Therefore, in order to use the Hückel method or the tight-binding approximation with a non-orthogonal basis it is advisable to pass first to an orthogonal basis by Löwdin's procedure (1950). Nevertheless, Eq. (5.2.6) can be used to estimate the difference between similar calculations with and without neglect of overlap, provided the overlap values are not too large. A study made on a linear chain of the type

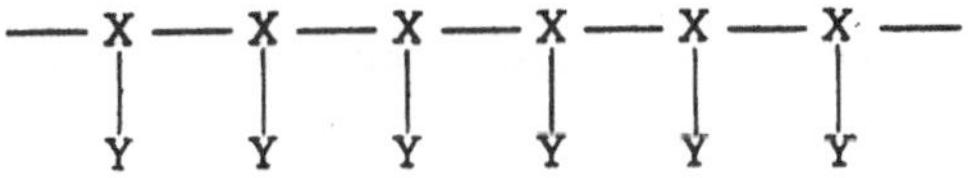

with four orbitals per unit cell showed marked differences in the band widths (and hence in the level densities) ; the consequences on properties like interaction with other molecules are quite serious, not to speak of the bulk properties normally related to band structure (Del Re *et al.* 1968; Tejeda and Shevchik 1976).

The non-orthogonality catastrophe can also be associated with correlation (Kaga and Yosida 1978).

5.2.3 Block-factorization of the overlap matrix. - As has been pointed out in Sec. 5.1.2, a factorization like that of Eq. 5.1.2 corresponds to localization only if the condition (5.1.12) is satisfied; in other words, if only orbitals participating in the same bond are allowed to overlap. Therefore, the overlap matrix must have the same block form as $\mathbf{H}$. On the other hand, it has been pointed out that a maximum localization procedure like that of Sec. 5.1.3 minimizes but does not suppress inter-bond overlaps, because that is not feasible by a unitary transformation not mixing orbitals of different centres. Therefore full inter-bond orthogonality must be obtained by allowing for some delocalization; the fact that inter-bond overlaps are generally small suggests that this will not greatly affect the physical significance of results. It can be carried out using a modification of Löwdin's transformation (Del Re 1973).

Let us call $\mathbf{S}$ the overlap matrix after hybridization, and $\mathbf{S}_0$ that part of it which corresponds to a block form of the type (5.1.2). Then a localized basis $|\underline{\chi}>_{loc}$ can be constructed from the hybridized basis $|\underline{\chi}>$ by the linear transformation

$$|\underline{\chi}>_{loc} = |\underline{\chi}> \sqrt{S^{-1}S_0} \qquad (5.2.6)$$

as can be seen by direct verification.

Although the square root looks rather formidable, it may be noted that $\mathbf{S} - \mathbf{S}_0$ is the matrix of the inter-bond overlaps, so that the argument of the square root is very close to the unit matrix; therefore, even the series expansion suggested by Löwdin (1956) should provide a satisfactory computational procedure. By the same token, we can expect the elements of $|\underline{\chi}>_{loc}$ to differ only slightly from the hybridized strictly atomic basis $|\underline{\chi}>$.

The above remark completes a very important argument in connection with model orbital schemes; it shows in what way an orbital basis corresponding to the familiar bond formula of a molecule can be constructed. In fact, the whole procedure finds out automatically lone pair, two-centre, and many-centre orbital subgroups, and prepares the basis for the corresponding model treatment.

5.3. General Orbital Bases

In the above considerations we have taken Eq.s (5.1.9) and (5.1.10) for granted. Actually they are a definition of the atomic orbitals used in an MO-LCMVAO method.

The very attribute 'atomic' is a very delicate one. Any complete basis of orbitals can be used to construct any other orbital, and a combination of, say, Slater orbitals of a given atom can give an orbital on *another* atom. Thus, the attribute in question must be related to a special Hamiltonian. That is why we have assumed in Eq.s (5.1.6) and (5.1.7) that it is possible to define an effective atomic Hamiltonian $\hat{H}_1^{eff}(X)$ which, by (5.1.10), is a hydrogen-like Hamiltonian - in the sense that it has the same degeneracies. The rest of the whole treatment follows from that premise. The Hamiltonian $\hat{H}_1^{eff}(X)$ may be called the Hamiltonian of the atom *in situ* (Mulliken 1949); and its eigenstates are the "modified atomic orbitals" (MAO) (Mulliken 1964, Del Re 1967).

We shall not stop on the definition of $\hat{H}_1^{eff}(X)$, for one thing because it is still a matter of discussion. The important point here is to have suggested how the ingredients of a simple semiempirical method should be discussed and defined, and shown some of the inconsistencies which can impair the scientific validity of such a method.

Still, a discussion of general bases is necessary. Even within the domain of orbital bases, we have to analyze the significance of truncation and linear transformations; we have to consider delocalized orbital bases; and we have to establish the connection between our considerations and the basis problem in 'ab initio' approaches.

5.3.1 Transformation of the basis and the EH method. - In order to connect the general theory of expansions into complete sets of states with the above considerations, consider a truncated basis of functions which are atomic orbitals in the sense specified above. In order to fulfill the condition that this basis could in principle be extended to give any desired accuracy to energy calculations, one must assume that the finite set of orbitals

$$|\mu\rangle = (|1A_1\rangle\ |2A_1\rangle \ldots |1A_2\rangle\ |2A_2\rangle \ldots) \qquad (5.3.1)$$

pertaining to the N atoms $A_1, A_2, \ldots, A_N$ of the given molecule belongs to a complete set $|\underline{\mu}\rangle_\infty$. The set $|\underline{\mu}\rangle$ is thus assumed to be formed by N subsets $|\underline{\mu}_j\rangle$ each relating to one atom A_j and formed by n_j elements.

Let us also introduce the sets $|\bar{\underline{\mu}}\rangle$ and $|\bar{\underline{\mu}}\rangle_\infty$ whose elements are Slater orbitals (or other atomic orbitals whose form is fully specified). We can obtain the basis $|\underline{\mu}\rangle$ by transformation in the Hilbert space spanned by $|\bar{\underline{\mu}}\rangle_\infty$ followed by (a) truncation of the basis, (b) a linear transformation, or viceversa.

It is easily verified that the two types of transformation are not equivalent in general. In fact if Π and Π_1 denote the matrices representing the projection operators which single out $|\underline{\mu}\rangle$ from $|\underline{\mu}\rangle_\infty$, and $|\bar{\underline{\mu}}\rangle$ from $|\underline{\mu}\rangle_\infty$, respectively; and if $\mathbf{T}$ and $\mathbf{T}_1$ are two linear transformations, the two cases correspond to

$$|\underline{\mu}\rangle = |\bar{\underline{\mu}}\rangle_\infty \mathbf{T}\Pi \qquad (5.3.2a)$$

$$|\underline{\mu}\rangle = |\bar{\underline{\mu}}\rangle_\infty \Pi_1 \mathbf{T}_1 \qquad (5.3.2b)$$

Therefore, unless $\mathbf{T}$ and $\mathbf{T}_1$ are related in a very special way, cases (a) and (b) are not equivalent. From the practical point of view, case (a) gives priority to the "choice of the form of the atomic orbitals," whereas case (b) relates to the problem "choice of preliminary combinations of atomic orbitals already forming a truncated set". Of course one can define a general transformation leading from $|\bar{\underline{\mu}}\rangle_\infty$ to $|\underline{\mu}\rangle$

$$|\underline{\mu}\rangle = |\bar{\underline{\mu}}\rangle_\infty \mathbf{T}\Pi\mathbf{T}_1 \qquad (5.3.3)$$

which can be read: We choose a complete set in the Hilbert space associated with the operator we are interested in; we single certain functions out of the whole complete set; we combine them linearly.

More precisely, we can think of our truncated atomic orbital set as being obtained by building atomic orbital of the types *s, p, d,* etc., which are linear combinations of Slater orbitals of the same species but with different principal quantum numbers, and by hybridizing them, i.e., combining them within subsets corresponding to individual atoms (Del Re 1967).

5.3.2. <u>The basis and the parameter problem.</u> - Once a pure MVAO basis has been prepared, it will be hybridized and/or subjected to linear transformations; current examples are symmetry adaptation and transformation to molecular orbitals of subsystems. Therefore, we can have Hamiltonians referred to: (a) MVAO's, (b) a model basis $|\underline{b}\rangle$ obtained by an *ad hoc* linear transformation, (c) the molecular orbital basis $|\underline{k}\rangle$.

The parameters may be chosen at stage (a) or (b). A widely held opinion is that it is sufficient to keep in mind that a transformation **T** transforms the Hamiltonian matrix **H** according to the rule

$$\mathbf{H} \rightarrow \mathbf{T}^{+}\mathbf{H}\mathbf{T} \tag{5.3.4}$$

As a matter of fact, the criteria adopted for the semi-empirical or approximate evaluation of the matrix elements of **H** are not independent of **T**, or, for that matter, of the zero point of energy. (A shift of the energy zero point by α_0 results when a matrix $\alpha_0\mathbf{S}$) is subtracted from **H**). This point was analyzed by Berthier *et al.* (1966) and partly taken into account by Pople and Segal (1965) in connection with invariance by rotation of two-electron integrals. We emphasize in particular that a parametrization carried out on a Hamiltonian referred to the basis $|\underline{\chi}>_{loc}$ of Eq. (5.2.6) by means of empirical relationships between overlap and transfer integrals will not give elements obeying the same relationships in the pure-MVAO basis $|\underline{\mu}>$, and vice versa. Such is the case with the Extended Hückel Method (EHT), whose parametrization involves the expression

$$H_{\mu\nu} = k<\mu|\nu> (H_{\mu\mu} + H_{\nu\nu})/2. \tag{5.3.5}$$

This expression is valid only in a specific basis, and is usually applied to the pure-MVAO basis: then it does not hold for the corresponding hybrids, even if, by virtue of the way in which they have been obtained, they are a basis physically and mathematically equivalent to $|\underline{\mu}>$. On the other hand, the simpler recipe assuming proportionality of overlap integrals to bond integrals is invariant under (5.3.4).

The above remark is especially important in view of the widespread use of EHT calculations for atom clusters in solid-state theory. For clusters of Ge-like atoms, for instance, where the *sp*-"band" may be expected to derive from tight-binding crystal orbitals on a basis of quasi-tetrahedral hybrids, it is not indifferent whether one assumes atomic hybrids or atomic *s* and *p* orbitals to obey Eq. (5.3.5).

5.4. π-Electron Models and the σ-π Separation

In chapter 3 and 4 we have considered pure π-electron and pure σ-electron models, after which we have discussed the so-called all-valence-electron models. The question of the validity of π-electron treatments and of separate treatments of the σ-core largely falls within the scope of the present chapter, as it involves considerations on the basis. The earliest treatment of the conditions for σ-π separation was published by Lykos and Parr (1956); that separation was extensively discussed by Kutzelnigg *et al.* (1971). We shall present here considerations which largely overlap those papers, but are recast in the language of the present study.

5.4.1. <u>π-Electrons or π-orbitals?</u> - The first remark to be made is that the very expression 'π-electrons' is incorrect, because electrons are identical. The question underlying the σ-π separation is better phrased as follows: Are there states of an unsaturated molecule that can be treated on a basis of π-orbitals? In mathematical language we are essentially inquiring whether there is a linear subspace (of the one-particle Hilbert space associated with the electrons of the given molecule) which breaks down into a space spanned by π atomic orbitals only and a space spanned by σ atomic orbitals only; so that an effective one-electron Hamiltonian matrix representing the given electron system in that subspace would be a block-diagonal matrix such as in the example (5.1.2).

In general, the separation of the states of a Hilbert space into subspaces does take place; indeed, the theory of spectroscopies like NMR rests precisely on such separations. In those widely accepted cases, however, the justification is very simple: The states to be discussed are almost degenerate, their separation in energy being several orders of magnitude smaller than the difference between their energy and that of any other state of the system.

In the case of σ and π electrons, such a justification is not acceptable *sic et simpliciter*. However, symmetry considerations may also be used to justify a separation; clearly, if all those elements of the given one-electron Hamiltonian which couple basis states of the two subspaces are zero by symmetry, or at least very small, then the Hamiltonian matrix is (exactly or almost exactly) block diagonal.

Physical considerations point toward a combination of the two effects in favor of the σ-π separation. As the one-electron Hamiltonian operator is totally symmetric, in planar molecules the matrix elements connecting π atomic orbitals, which are antisymmetric with respect to the place of the molecule, to σ atomic orbitals are rigorously zero; in locally planar molecules they should be expected to be small, because only distant σ atomic orbitals would couple to a given π atomic orbital.

Moreover, the π atomic orbitals of the various atoms of a molecule are either degenerate or closer in energy to one another than to σ atomic orbitals (cf. Table 1.3): therefore, they must be expected to mix among them much more than with σ orbitals.

5.4.2. Many-electron aspects of the σ-π separation. - So far we have been concerned with orbitals. What about the many-electron states? If the basis is such that

$H_{\mu\nu} \approx 0$ if μ is a σ-AO, and ν a π-AO or vice versa

$$(\mu\nu|\rho\tau) \cong 0 \text{ if } \mu \text{ is a } \sigma\text{-AO, and } \nu \text{ a } \pi\text{-AO or vice versa, and/or } \rho \text{ is a } \sigma\text{-AO, and } \tau \text{ a } \pi\text{-AO or vice versa.} \quad (5.4.1)$$

then, using the letter a for creation and annihilation operators of π orbitals, the letter b for those of σ orbitals, the Hamiltonian can be written

$$\hat{H} \cong R_0 + \sum_{\mu,\nu,s}^{(\sigma)} H_{\mu\nu}^{(\sigma)} \hat{b}^+_{\mu s}\hat{b}_{\nu s} + \sum_{\mu,\nu,s}^{(\pi)} H_{\mu\nu}^{(\pi)} \hat{a}^+_{\mu s}\hat{a}_{\nu s} + \frac{1}{2}\sum_{\mu,\nu,s}^{(\sigma)} \sum_{\rho,\tau,s'}^{(\sigma)} (\mu\nu|\rho\tau)\hat{b}^+_{\mu s}\hat{b}^+_{\rho s'}\hat{b}_{\tau s'}\hat{b}_{\nu s} + \frac{1}{2}\sum_{\mu,\nu,s}^{(\pi)} \sum_{\rho,\tau,s'}^{(\pi)} (\mu\nu|\rho\tau) \hat{a}^+_{\mu s}\hat{a}^+_{\rho s'}\hat{a}_{\tau s'}\hat{a}_{\nu s'} + \sum_{\mu,\nu,s}^{(\pi)} \sum_{\rho,\tau,s'}^{(\sigma)} (\mu\nu|\rho\tau)\hat{a}^+_{\mu s}\hat{b}^+_{\rho s'}\hat{b}_{\tau s'}\hat{a}_{\nu s}. \quad (5.4.2)$$

A partially SCF form of this can be obtained by keeping in mind that

$$\hat{a}^+_{\mu s}\hat{b}^+_{\rho s'}\hat{b}_{\tau s'}\hat{a}_{\nu s} = \hat{b}^+_{\rho s'}\hat{b}_{\tau s'}\hat{a}^+_{\mu s}\hat{a}_{\nu s}; \quad (5.1.3)$$

then we can replace $\hat{b}^+_{\rho s'}\hat{b}_{\tau s'}$ by its expectation value over the σ state, $<\hat{b}^+_{\rho s'}\hat{b}_{\tau s'}>$, so that

$$\hat{H} \cong R_0 + \hat{H}^{(\sigma)} + \hat{H}^{(\pi)} \quad (5.4.4)$$

where

$$\hat{H}^{(\pi)} = \sum_{\mu,\nu,s}^{(\pi)} \left(H_{\mu\nu}^{(\pi)} + \sum_{\rho,\tau,s'}^{(\sigma)} (\mu\nu|\rho\tau)<\hat{b}^+_{\rho s'}\hat{b}_{\tau s'}>\right)\hat{a}^+_{\mu s}\hat{a}_{\nu s} + \frac{1}{2}\sum_{\mu,\nu,s}^{(\pi)} \sum_{\rho,\tau,s'}^{(\pi)} (\mu\nu|\rho\tau)\hat{a}^+_{\mu s}\hat{a}^+_{\rho s'}\hat{a}_{\tau s'}\hat{a}_{\nu s}. \quad (5.4.5)$$

This defines a π many-electron Hamiltonian, where the π electrons just contribute to the one-electron potential. Of course $\hat{H}^{(\sigma)}$ is defined as the difference between $\hat{H}$ and $\hat{H}^{(\pi)}$. The important conceptual conclusion is that $\hat{H}$ is now divided into an operator which acts only on the σ orbitals entering the given many-electron state, and one which acts only on its π orbitals. Thus, we can think of a set of states which consist of a given (even unknown) eigenstate of $H^{(\sigma)}$ and of the various eigenstates of $H^{(\pi)}$

that can be obtained from the given basis. Of course, knowledge of the density matrix $\hat{b}^+_{\rho s}$, $\hat{b}_{\tau s}$, is necessary if the 'best' separation is desired; but a model may be introduced by assigning the σ density matrix a standard form, i.e. by assuming that the σ frame is not affected by changes in the π system.

The validity of π models thus rests on Eq. 5 (5.4.1) and on the SCF Ansatz in the π-σ two-electron term. The former are generally good because of the symmetry reasons mentioned in the preceding subsection; the latter has the shortcomings of current SCF assumptions, but is at least as reasonable as most SCF calculations.

A π-one electron Hamiltonian can be obtained from (5.4,5) in the usual way. Clearly all the approximations of the PPP type are admitted.

5.4.3. An application of π-electron schemes: Azulene. - Azulene (I) is an unsaturated hydrocarbon $C_{10}H_8$ which results from the fusion of a pentagonal and a heptagonal cycle. Its properties are markedly different from those of its benzene isomer, naphthalene (II)

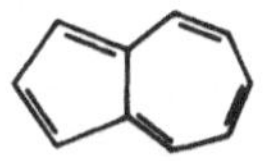

Ia

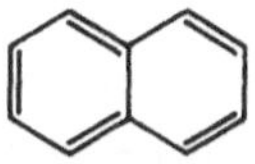

IIa

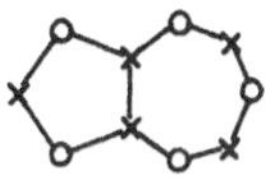

Ib

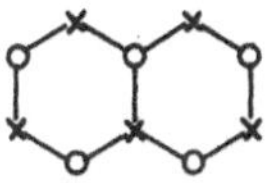

IIb

It exhibits a blue colour, it is a good triplet-state quencher, and has quite an appreciable dipole moment for a hydrocarbon: 1 D according to Wheland and Mann (1949).

Azulene belongs to the class of conjugated hydrocarbons called non-alternant by Coulson and Rushbrooke (1940). Alternance and non-alternance, as has been already recalled in sec. 2.1.3., is a topological property of the σ-frame formed by the CC bonds of the molecule. Using the language of coloured graphs (Graovac *et al.* 1977), it may be said that naphthalene, like any benzenoid hydrocarbon, can be represented by a two-colour graph (II*b*), whereas the azulene frame cannot, because of the presence of odd carbon-atom cycles (I*b*). In the Hückel model, the topological characteristics of the σ frame have important consequences on the distribution of the π electrons within a conjugated hydrocarbon, because of the nearest neighbour approximation (bond integral $\beta_{\mu\nu}$ different from zero only when $|\mu\rangle$ and $|\nu\rangle$ are π-orbitals of atoms linked by a σ bond).

Let us refer here to the iterative (SCF) Hückel method, in which the atomic parameter α_μ depends on the π charge carried by the atom M, so that all the α's of a conjugated hydrocarbon may be assumed to be equal if the carbon atoms carry the same charge (e.g. $q_M = 1$).

Coulson and Rushbrooke (1940; and many others after them in a more sophisticated way) proved that in the Hückel method the π charge of every carbon atom of an alternant hydrocarbon is unity; so that the approximation consisting in taking equal α's corresponds to self-consistency, and the molecule has no permanent electric dipole $\vec{\mu}$. *A contrario* non-alternant hydrocarbons like azulene are expected to exhibit significant intramolecular π-electron transfers for equal α's; adoption of an iterative scheme (and hence of different α values) is necessary in order to evaluate correctly that quantity (Berthier and B. Pullman 1949).

In fact, the first calculations on azulene, carried out by Coulson and Longuet-Higgins (1947) suggested the existence of an electric dipole the negative end of which lies in the five-membered cycle, the positive end in the seven-membered cycle; the magnitude obtained was very large: 6.4 D. Still within the Hückel model, A. Pullman and Berthier (1949) identified two sources of error in that treatment. 1) In a non-alternant hydrocarbon, the π charges depend on the values assigned to the bond integrals β_{CC}, even if the same Coulomb integral is assigned to each carbon atom. Now, in azulene, the CC bond *b* common to the two rings is far longer than the peripheral bonds. Therefore, that bond could be assigned a bond integral smaller in absolute value, and the dipole moment should decrease, for the situation would get closer to an alternant ten-membered cycle. With $\beta_b = 0.8\ \beta_0$, one obtains $|\vec{\mu}| = 5.25$ D. 2) Following Wheland and Mann (1949), and in accordance with sec. (4.4), the variation of the Coulomb integral α_M can be assumed to be a linear function of the net charge of M, defined according to Eq.s (2.1.22) and (6.2.6) as $q_M = 1 - p_M$:

$$\alpha_M = \alpha_0 + kq_M\beta_0 \quad (k > 0)$$

Then a sort of feedback effect takes place and the required iterative calculation gives $|\vec{\mu}| = 3.8$ D with k = 1.

The indications of the classical Hückel model have not been modified by subsequent calculations on azulene (except with respect to the numerical values of π charges) at least as concerns the orientation of the dipole moment. Calculations by Julg (1955; $|\mu| = 1.7$ D) and by Pariser (1956) at the π-SCF and at the π-SCF + CI level, respectively, confirm the theoretical prediction of an important ground-state polarity of azulene. The *ab initio* calculations carried out so far (Buenker and Peyerimhoff 1969; Christoffersen 1971) have not altered his picture very much.

5.5. The Basis Problem in Solid-state Physics.

The fact that the choice of the basis is not a purely mathematical question becomes especially evident in solid state theory, where systems are found containing both highly localized and highly delocalized electrons. Now, the AO scheme corresponds to a limiting case where an electron is essentially localized on the various atoms, the coupling between the atomic states being very small. In ideal metals, the opposite limiting case is approached, namely the states of an electron in a uniform potential. As is well known, those states are plane waves, and we can thus speak of a plane-wave-orbital (PWO) basis; depending on the boundary conditions, it can be a discrete or a continuous basis.

The two basic approaches of solid-state physics, the tight-binding (TB) and the plane-wave (PW) scheme, are characterized by the two types of bases. We shall discuss here the reasons why the areas of application are different and the ways in which the two types of bases can be combined. We shall use the notation:

$$|\underline{k}> \text{ for the PW orbital basis,}$$
$$|\underline{\mu}> \text{ for the AO (TB) basis.} \quad (5.5.1)$$

5.5.1. Completeness and equivalence. - In principle, both the PWO basis and the AO basis are complete; indeed, the AO basis is overcomplete if all the orbitals of all the atoms are included in it, for the orbitals of a single atom already form a complete set. This means that any given one-electron state can be reproduced equally well by a linear combination of PWO and by a linear combination of one-centre AO's. This is what is meant by saying that two complete bases in the same Hilbert space are equivalent.

Mathematical equivalence is not physical equivalence. This can be easily seen by comparing the representations of an atomic orbital (say, a 2s one) of atom B in terms of a one-centre AO basis on atom A, an ordinary AO basis on atom B, a PWO basis. In the former case, it is necessary to superpose orbitals of A in such a way that they will cancel on A itself and everywhere else except in the neighbourhood of a point (B) located at a certain distance from A; such a result is possible, but it will demand very high quantum numbers (and possibly the continuum states), in other words a very large number of basis functions. On the other hand, if the many-centre AO basis contains precisely the orbital of B under consideration, just one basis state will be sufficient in the second case. The third case, if the basis is continuous, will demand a Fourier integral involving practically all the plane waves of the basis; there will have to be destructive interference outside the region of the orbital of B under consideration, constructive interference inside that region. However, the situation will not be as bad as in the first case, because the basis is uniformly distributed, and not concentrated in a small region of space.

In conclusion, the choice of the basis is extremely important to ensure rapid convergence at least for the states under study. It becomes crucial, of course, when the basis is truncated, because then the elements retained must reproduce as well as possible the states one is interested in. This consideration is a physical one because in a given class of bases there is an optimal choice suitable for a given type of physical problem, just as there is an optimal choice of the reference system used to describe the motion of the planets around the sun. (Even in *ab initio* schemes of, say, the double-zeta type there exists an optimum choice of orbital exponents, which can be treated as the physically correct choice.)

A numerical example can be used to illustrate the above considerations. Let us consider a one-dimensional problem characterized by a variable x such that $-L \le x \le L$. Then two complete bases are given by the sets of functions

$$\chi_1 = -\sqrt{3/2}, \qquad \chi_n = (-1)^{n+1}\sqrt{(2n+1)/2} \cdot P_n(x/L) \tag{5.5.1}$$

$$k_1 = 1/\sqrt{2\pi}, \qquad k_n = \frac{1}{\sqrt{\pi}} \cos\left\{\left[1 + (-1)^n\right]\frac{\pi}{4} - \frac{n\pi x}{2L}\right\}; \tag{5.5.2}$$

where P_n is the n-th Legendre polynomial. It is easy to check that the qualitative behaviour of the two sets of functions is quite different especially at the extremes of the given interval. For instance, χ_2 is tangent to k_2 for $x = 0$, but for $x = L$ it is $\sqrt{5/2}$ whereas k_2 vanishes. The point we are discussing is illustrated by a state represented by the function χ_2. A trivial exercise shows that in order to reproduce that function by the Fourier bases (5.5.2) it is necessary to combine all the sine functions of the latter:

$$\chi_2 = \sqrt{5/2} \cdot (1/L) \sum_{\nu} \frac{(-1)^{\nu}}{\nu} \sin \nu\pi\, x/L. \tag{5.5.3}$$

Thus, the Fourier basis will no longer be equivalent to the Legendre basis if it is truncated so as to retain only a few sine functions; there will be no equivalence at all if the functions with n even are suppressed, for it will be impossible in that case to reproduce odd-degree polynomials.

5.5.2. Orthogonalized plane waves (OPW) method. - In real crystals the inner electrons of atoms may be supposed to be tightly bound to their centres, whereas the valence electrons are more or less delocalized. Therefore, even in a one electron scheme, it is desirable to have a basis combining the features of an AO basis (for the low-energy orbitals) with those of a PW basis (for mobile electrons). There are two ways of approaching the question. The first one consists in deciding that the localized electrons are excluded from the model treatement to be carried out. Then, attention is focused on the PW basis, but something must be done to make it impossible for a variational procedure - which is the procedure corresponding to solution of an eigenvalue equation - to produce just poor approximations to strongly bound localized orbitals. In other words, we want a single basis which should be in part a many-centre AO basis, in part a PW one, the former part being used to describe inner electrons, the latter to describe delocalized electrons from the valence shells of the atoms under study.

In practice, one may proceed as follows. First of all, the symmetry adapted (Bloch) orbitals are constructed from the inner AO's $|\nu>$: the general term of the new basis $|\underline{\chi}(k)>$ is

$$|\chi(\vec{\underline{k}})> = \sum_{\nu} \exp(i\vec{\underline{k}} \cdot \vec{\underline{R}}_{\nu})|\nu> \qquad (5.5.4)$$

where $\vec{\underline{R}}_{\nu}$ is the position vector of the centre of $|\nu>$. We now suppose that there are certain crystal orbitals, namely the deep-lying orbitals, which can be described correctly by the basis (5.5.4). In order to describe the other orbitals we want to use the PWO basis $|\underline{k}>$. Such a basis consists of states $|\vec{\underline{k}} + \vec{\underline{K}}_n>$ represented by wave functions of the type

$$\exp(-i(\vec{\underline{k}} + \vec{\underline{K}}_n) \cdot \vec{\underline{r}})$$

- $\vec{\underline{K}}_n$ being a vector of the reciprocal lattice, $\vec{\underline{r}}$ being the position vector of the electron - ; however, as has been mentioned, it cannot be used as such, because it may well end up by giving poor approximations to states already described by the basis $|\underline{\chi}(\vec{\underline{k}})>$. Because of the variational theorem, if the basis $|\underline{k}>$ has in any way the possibility of approximately reproducing a Bloch orbital already described by the basis (5.5.4), it will. However, being truncated and not rapidly converging to localized states, it will do so poorly. What is worse is that the other crystal orbitals obtained from the basis $|k>$ will consequently be much worse approximations to the exact ones than they would otherwise. Therefore, following Herring (1940), it is advisable to replace the PWO basis by an OPWO (orthogonalized-plane-wave orbital) basis which is obtained from the former by eliminating the part that could reproduce the orbitals described by (5.5.4). This is done by orthogonalization: the basis $|\underline{k}_{OPW}>$ is required to satisfy the condition

$$<\underline{k}_{OPW}|\chi(\vec{\underline{k}})> = 0. \qquad (5.5.5)$$

In order to see how this condition can be satisfied, suppose the basis $|\underline{\chi}(\underline{k})>$ contains p elements, and the basis $|\underline{k}>$ contains q elements. We define our OPWO basis by introducing a $p \times q$ matrix T such that

$$|\underline{k}_{OPW}> = |\underline{k}> - |\underline{\chi}(k)> \mathbf{T}, \tag{5.5.6}$$

under the condition (5.5.5). The latter becomes a linear system of equations for the elements of **T**:

$$< \chi(\underline{k}')|\underline{k} + \underline{K}_n> - \sum_{\nu} <\chi(k')|\chi(k)> T_{\nu n} = 0. \tag{5.5.7}$$

It must be emphasized that (5.5.6) does not represent the only possibility for defining the OPWO basis. For instance, one could construct it from a PWO basis containing $p + q$ elements, and contract it so as to finally obtain a basis containing q terms consisting of superpositions of plane waves. The actual choice depends on considerations of numerical convergency and of physical significance - i.e. on the nature of the states to be described by the OPWO basis. As far as we know, only the Ansatz (5.5.6) has been actually used (cf. Rössler and Treusch 1972).

5.5.3. The 'augmented-plane-wave' (APW) method, the X_α method, and related approaches. - The OPWO basis is used to separate strongly localized orbitals from the others; we have pointed out that it need not even contain contributions from localized orbitals, even though it normally does. A different point of view consists in dividing space into different regions where a given orbital has different properties, and to construct a basis whose elements closely approach those properties in the various regions. For instance, one may try do construct a basis which is close to atomic orbitals in the neighbourhood of nuclei and is close to plane wave elsewhere; this is the idea developed by Slater (1937) in the APW method. The "atomic" regions are spheres around the atoms; there the basis states are supposed to satisfy the atomic Schrödinger equation; but they vanish abruptly outside those spheres. Other basis states are plane waves outside the atomic spheres, and vanish inside them. Although the method could be formulated in terms of the above basis, in practice everything goes as if the basis itself depended on the energy of the state to be determined. In this version, the method has been applied successfullly both in solid-state physics and in quantum chemistry, where it seems to render good services especially for heavy atoms. Of course, application to molecules demands an *ad hoc* elaboration of the basic idea; and the result is called the X_α method (Johnson 1973).

We shall not linger at the problem of the bases underlying such special methods, because that is a subject for future research; but we point out that methods like X_α

emphasize that the interesting question is really the choice of a trial function, which does not always imply the choice of a standard basis. We present in the next section a very topical problem where the X_α method has proved to be particularly useful.

5.6. Metal-metal Bonds and the X_α Method.

5.6.1. Metal-metal bond. - It is now established (Vahrenkamp 1978, Huheey 1972) that all metallic elements participate in metal-metal bonding and that bonds between d elements are particularly strong. For example the lightest metal Li and the heaviest one U give metal-metal bonds; the first one in $[Li\ CH_3]_4$ and the second one in $U_6O_4(OH)_4(SO_4)_6$. The transition elements in the middle of the transition metal series are the elements giving the largest number of compounds with metal-metal bonds.

Usually, if the metal achieves an 18-electron configuration by covalent bonding with other metals, it is clear that there is a metal-metal bond. But, in many other cases, the metal-metal bond is due to the absence of other bonding partners or to structural constraints such as the ones due to bridging ligands.

The metal-metal bonds are found in units containing two metal atoms, in aggregates such as rings, cages and cluster compounds (not to be confused with "clusters" as defined by physicists - i.e. aggregates of equal atoms), and in extended arrangements such as chain structures, layers and three-dimensional networks.

In two-metal units there are single bonds and multiple bonds. Single bonds are attributed when the two fragments need one electron to complete the inert gas configuration. Bond orders of 2, 3 and 4 between metal atoms are now frequent. The elements showing the greatest tendency to form multiple bonds are chromium, molybdenum, tungsten and rhenium. It seems that these units are more stable if there are bridging ligands protecting the metals. The H atoms can play such a role and up to four may occur as in $H_8Re_2(P\ Et_2\ Ph)_4$ (Bau *et al.* 1977) of which the structure has been determined and is given in Fig. 1. From electron considerations, the Re-Re bond can be termed formally a triple bond.

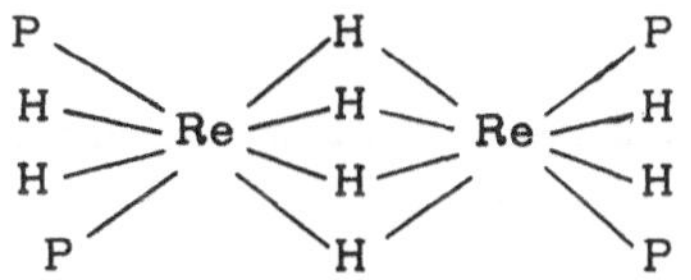

P is for P Et_2 Ph

Fig. 1

It is difficult to predict whether aggregates or simple systems with metal-metal multiple bonds will be formed. But it can be said that large numbers of ligands or anions favor multiple bonding, small numbers promote aggregation. Three and four metal atoms happen to be joined together in rings containing four to eight members. An example is given in Fig. 2 (Ernst et al. 1977):

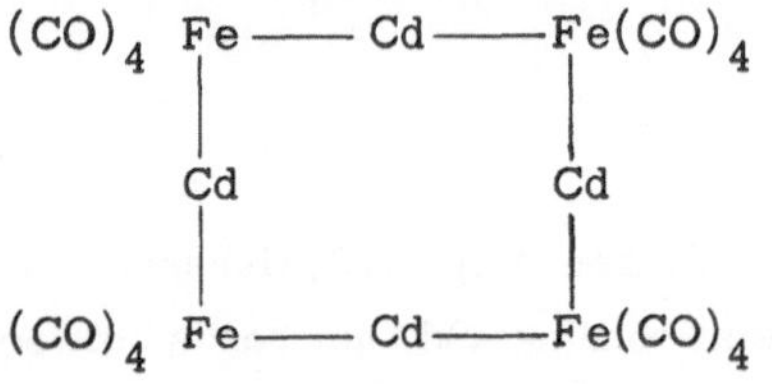

Fig. 2

Main group metals give clusters as well as transition metals. For example the cations Bi_8^{2+} and Bi_9^{5+} are well known (Corbett 1976). Clusters of six molybdenum, niobium and tantalum have been synthesized. In $[Mo_6Cl_8]Cl_4$ each molybdenum atom is surrounded by four chlorine atoms above and four molybdenum atoms below (Huheey 1972).

In the chain structures there are square - planar d^8 complexes which are stacked with short metal-metal bonds as in $[Ir(CO)_3Cl]_x$ (Reis *et al.* 1977), where each iridium atom achieves an 18-electron configuration by two metal-metal bonds of length 2.844 Å. Optical, electrical, magnetic and structural properties due to a one-dimensional infinite linkage of metal atoms are found after oxidation of tetracyanoplatinate ions (Miller and Epstein 1976). A variety of complexes have been prepared; one example is

$$K_2\,Pt(CN)_4\,X_{\sim 0,3}\,(H_2O)_x \qquad (X = Cl \text{ or } Br)$$

with all platinum atoms in the equivalent oxidation state of ~ 2.3.

Two - and three dimensional metal-metal interactions exist also. It is the case, for example, in the monochlorides of the lanthanoids, Gd Cl and Tb Cl (Simon *et al.* 1976), where double layers of metal atoms M alternate with double layers of non-metal atoms X (XMMX ... XMMX ...).

5.6.2. Bond Lengths. - It is generally accepted that a metal-metal bond exists when the metal atom distances are of the same order of magnitude as in the metal.

For example, in $Re_2Cl_8^{2-}$, the Re-Re distance is rather short: 2.24 Å compared with a Re-Re distance of 2.75 Å in rhenium metal and 2.48 Å in Re_3Cl_9 (Huheey 1972).

The identification of a metal-metal bond requires the comparative study of two compounds of similar stereochemistry. For example (Vahrenkamp 1978), the change of bond length between the compounds (1) and (2) is a proof that there is no metal-metal bond in (1) and a metal-metal bond in (2).

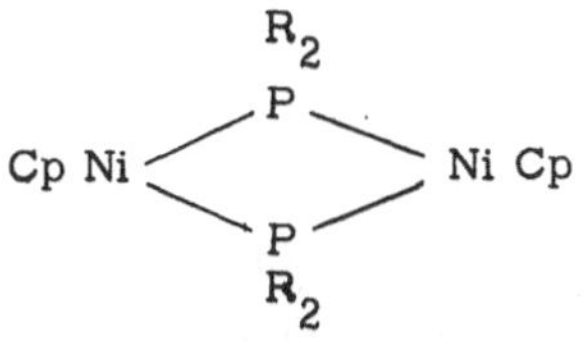

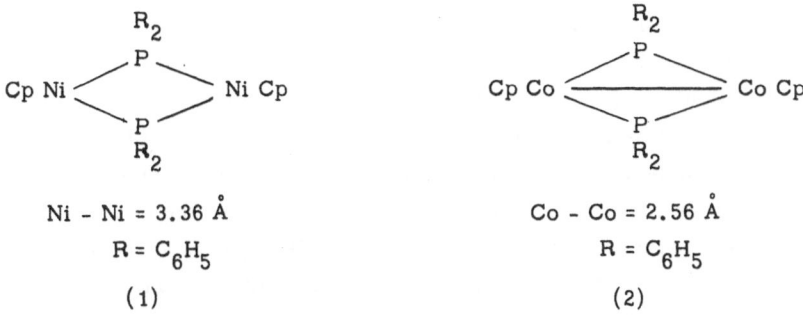

Ni - Ni = 3.36 Å

R = C_6H_5

(1)

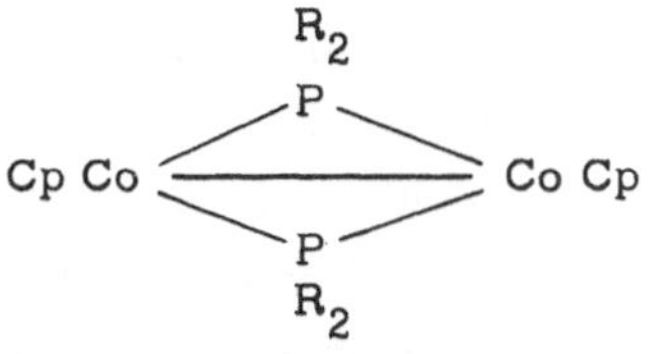

Co - Co = 2.56 Å

R = C_6H_5

(2)

It is possible to see by the study of the bond lengths that the strength of the bond between the metal atoms increases with increasing atomic weight.

The complexes (3) demonstrate this fact (Vahrenkamp 1978).

$[Cl_3M - (\mu - Cl)_3 - MCl_3]^{3-}$

M = Cr, Mo, W

(3)

Cr - Cr = 3.12 Å

Mo - Mo = 2.66 Å

W - W = 2.41 Å

Furthermore, it is found by studying the Cr - Cr distances among the Cr-carboxylato compounds that the Cr - Cr distance shows a tendency to increase as the distance between the chromium atom Cr and the axial ligand L_{ax}, Cr - L_{ax}, decreases. For these compounds, it is usually recognized that there is a quadruple bond between the metal atoms. While quadruple Mo - Mo and Re - Re bonds are each restricted to a narrow range of lengths (~ 0.1.Å), quadruple Cr - Cr bonds cover an enormous range, from 2.541 Å in $Cr_2(O_2CCF_3)_4(Et_2O)_2$ to 1.847 Å in $Cr_2[2,6-C_6H_3(OMe)_2]_4$ (Cotton 1978).

By the determination of the bond lengths in $H_8Re_2(PEt_2Ph)_4$ it was also possible to say that the Re - Re bond could be called formally a triple bond. Indeed the Re - Re distance in $H_8Re_2(PEt_2Ph)_4$ [2.538 Å] is intermediate between that of the classic quadruply bonded anion $[Re_2Cl_8]^{2-}$ [2.241 Å] and the formally double-bonded compound $H_2Re_2(CO)_8$ [2.896 Å] (Bau *et al.* 1977).

Magnetism. - Magnetic exchange interactions between paramagnetic transition metal ions are characterized by coupling of spins between two or more metal ions, giving rise to magnetic properties which differ substantially from those expected for a series of non interacting paramagnetic centres (Doedens 1976). Exchange interactions are classified as ferromagnetic or anti-ferromagnetic depending on whether cooperative alignment or anti-alignment of spins predominates. Antiferromagnetism is far more common but ferromagnetism occurs in the case of CrO_2, for example, where there is little possibility of metal-metal bonding (Baird 1968).

Bulk magnetic susceptibility measurements are the principal means by which metal-metal interactions are studied. Magnetic measurements over a range of temperatures are very useful particularly in giving information about the nature of the interaction in a particular compound by use of the Heisenberg-Dirac-Van Vleck (HDVV) model. In this model, under certain assumptions, one may derive an expression for the magnetic susceptibility as a function of various parameters including the temperature, the g values of the metal ions, and the exchange integrals J_{ij} between each pair of interacting ions. Usually the molar paramagnetic susceptibility versus T data are fitted to the theoretical expression, using the J and g values as parameters.

For example, magnetic susceptibility measurements indicate the presence of Cu - Cu interaction in the tetranuclear $[Cu_2OH(O_2CCF_3)_3(\text{quinoline})_2]_2$ molecules (Doedens 1976): All four metal ions must be interacting magnetically and at least a portion of the interaction must be transmitted via the bridging hydroxide ion.

The trifluoroacetate dimer, $[Cu(O_2CCF_3)_2(\text{quinoline})_2]_2$, and the corresponding acetate were also studied. The trifluoroacetate dimer has magnetic properties very similar to those of copper acetate itself, the relative base strengths of the bridging caboxylate anions are not major factors in the Cu - Cu interaction. Also the large difference in Cu - Cu separation between the magnetically similar acetate and trifluoroacetate adducts shows that the metal-metal distance is not a major factor in determining the strength of the interaction (Doedens 1976).

The chromium (II) carboxylates were the first antiferromagnetic carboxylates studied and many more chromium (II) complexes with formula $Cr\,L(R\,CO_2)_2$ have now been shown to have magnetic moments less than $1\,\mu_B$. The dimeric carboxylates of the second - and third - row transition elements are very strongly coupled and are interpreted as showing direct and even multiple metal-metal bonding (Catterick and Thornton 1977).

Bond energies. - All the values known until now are very uncertain. For example, determinations by way of mass spectrometric appearance potentials or recombination kinetics after flash photolysis gave 25 and 36 kcal. $mole^{-1}$ for D(Mn - Mn) in

$Mn_2(CO)_{10}$. By the determination of the activation energy of a reaction proceeding via cleavage of the metal-metal bond 37 kcal. $mole^{-1}$ was found for the same bond.

But all the measurements on bond energies confirm the fact that bonds between metals of the first transition series are extremely weak while the bonds between heavier transition metals are very strong. For example, an extrapolation using Raman data on overtones of the $M_2X_8^{n-}$ species give dissociation energies in the ranges 110-160 and 115-130 kcal. $mole^{-1}$ for the quadruple Mo - Mo and Re - Re bonds (Trogler *et al.* 1977).

Both theoretical calculations and experimental spectroscopic studies on complexes $M_2X_8^{2-}$ are consistent with the idea that the HOMO and the LUMO are orbitals of type δ and δ^* and that these orbitals possess very little ligand character (Trogler and Gray 1978). In complexes where these orbitals contain an odd number of electrons, the $\delta \rightarrow \delta^*$ vertical excitation energy is roughly equal to two times the δ bond dissociation energy.

Thus upper limits of 20 and 18 kcal/mole respectively are obtained for the δ bond dissociation energies in $Mo_2(SO_4)_4^{3-}$ (Erwin *et al.* 1977) and $Tc_2Cl_8^{3-}$ (Cotton *et al.* 1977).

Vibrational spectra. - For metal-metal single bonds in dinuclear complexes there are vibrational frequencies of 150-250 cm^{-1}, corresponding to force constants of 0.5 - 1.5 mdyn/ Å. The force constants increase on going from the lighter to the heavier metals.

A force constant of 2.7 mdyn/Å has been found for a Fe - Fe double bond and a range of values of 3.5 - 4.5 mdyn/Å for the Mo - Mo and the Re - Re triple and quadruple bonds.

In the case of clusters the assignment of vibrational frequencies is difficult.

Preparation and Chemistry (Vahrenkamp 1978). - There is too large a number of types of compounds for presenting some general rules for their preparation. In general the metal must be in a low oxidation state and the preparation is easier with heavy transition metals. If the mononuclear species has an odd number of electrons, dimerization to form a metal-metal bond becomes likely.

The majority of complexes containing more than one transition metal are prepared by a metathetic reaction of a metal-halogen compound with an anionic metal complex (Abel and Stone 1970):

$$Na\,Mn(CO)_5 + Re(CO)_5Cl \rightarrow (OC)_5\,Mn\,Re(CO)_5 + NaCl$$

By elimination of bonding partners from mononuclear compounds, it is possible to prepare M_2 units. Enneacarbonyldy-iron is prepared by irradiating pentacarbonyl-iron

$$2\,Fe(CO)_5 \xrightarrow{h\nu} Fe_2(CO)_9 + CO$$

It is also possible to obtain osmium carbonyl clusters containing up to eight osmium atoms by pyrolysis of $Os_3(CO)_{12}$ (Eady *et al.* 1975).

Bridging atoms can facilitate the formation of metal-metal bonds; for example, many carbonyl clusters have been prepared in which a sulfur is symmetrically bridged to a Co_3 triangle to form a tetrahedral grouping (Penfold and Robinson 1973).

Transition metal cluster compounds may be regarded as species intermediate between simple monometallic covalently bonded compounds and the free metals. Thus metal cluster compounds have colours more intense than those of comparable transition metal derivatives without any metal-metal bonding. This is exemplified by the green-black colour of $Fe_3(CO)_{12}$ and the black colour of $Co_4(CO)_{12}$ (King 1972). Outside reactions of rearrangements and ligand substitution, there are reactions directly affecting the metal-metal bond. The polynuclear complexes are very sensitive towards nucleophiles. The breaking of the metal-metal bond is promoted by the fact that the metal-ligand bonds are more energetic than the metal-metal bonds. For example, hard bases, acetylenes and even CO cleave metal-metal bonds in carbonyl complexes of light transition metals (Vahrenkamp 1978). In many cases, nucleophilic attack is preceded by rupture of the metal-metal bond. Even the quadruple bond of M_2X_8 compounds of molybdenum and rhenium is broken in presence of suitable ligands.

Addition is possible on triple bonds without breaking them. For example, there are a number of reactions such as

$$Cp_2M_2(CO)_4 + I_2 \rightarrow Cp_2M_2(CO)_4I_2$$

in which there is addition of a substrate across a M - M triple bond to yield a compound containing a M - M double bond (Klingler *et al.* 1975). M - M triple bonds react with small unsaturated molecules to give compounds with an increased M - M bond length. For example (Bailey Jr. *et al.* 1978), $Cp_2Mo_2(CO)_4$ reacts in solution with allene to give a compound where the four-electron donor ligand spans the Mo - Mo bond and where the Mo - Mo distance increases in length from 2.40 Å in $Cp_2Mo_2(CO)_4$ to 3.117 Å in $Cp_2Mo_2(CO)_4C_3H_4$. But $Mo_2(O\,Pr\text{-}i)_6$ reacts with nitric oxide to yield $[Mo(O\,Pr\text{-}i)_3NO]_2$ where a Mo - Mo separation of 3.335 Å probably signifies an absence of metal-to-metal bonding (Chisholm *et al.* 1978).

The ability of compounds containing M - M bonds to enter in reactions with nucleophilic compounds suggests that they may be active as catalysts. In fact, several examples are known (Vahrenkamp 1978; Chisholm and Cotton 1978). Thus it has been discovered that $Ru_3(CO_{12})$ catalyzes the water gas shift reaction (Laine et al. 1977).

$$CO + H_2O \rightleftarrows CO_2 + H_2$$

at low temperature, and $Ir_4(CO)_{12}$ the Fischer-Tropsch synthesis (Demitras and Muetterties 1977)

$$n\,CO + (2n+1)H_2 \rightarrow C_nH_{2n+2} + n\,H_2O$$

in which the hydrocarbons were methane, ethane, propane and isobutane, the latter two being minor constituents.

Clearly it can be said now that the chemistry of the metal-metal bond is very rich and likely the compounds with such bonds will find utilizations in the near future.

5.6.3. Theoretical treatments of metal-metal bonds. - Around the mid-sixties the theory was challenged with the explanation of the metal-metal bond. The facts to explain were the geometry of this new kind of complexes, the M - M bonding tendency in compounds other than the carbonyl types to be greatest among those second and third transition series metals lying to the left, the stability of the M - M bonds only where metal atoms are in low formal oxidation states (Cotton 1969).

The first attempts to this explanation were based on the 18-electron rule since systems with and odd-number of electrons are rare and since only heavy and late transition metals form 16-electron complexes. This electron-counting method is also used when is considered the close relationship between carbonyl cluster compounds and borane cluster compounds and when pairs of non-bonding electrons are allocated to each metal atom. (Kettle 1965). Based on the simple MO method (Wade 1976; Crossman *et al.* 1963), there were bond energy estimates with the aid of the symmetry properties of the metal atomic orbitals and of group theory. Usually the metal-metal and the metal-ligand bonds were separately considered.

By quantitative semi-empirical MO procedures and particularly by the extended Hückel method it became possible to explain the magnetic interactions (Hay *et al.* 1975) or to develop an interpretation of the electronic spectra and to assign the bands to precise transitions (Levenson and Gray 1975). But the EHT method applied to such molecules implies many parameters. At the beginning of the seventies it seemed difficult to perform ab initio calculations on the complexes with M - M bonds because these molecules are very large. Finally, the so-called X_α method usually applied to metals and alloys, was tested on these problems (Cotton 1978).

The essential features of the X_α method are summarized by the remarks of sec. 5.5.3 (Connolly 1977; Johnson 1973, 1975) and by the further remark that the kinetic energy is treated as in the Hartree-Fock model but the exchange term is obtained by a statistical approximation.

One starts with a general density matrix formulation of the many-electron problem. If the Hamiltonian **H** of the system is expressed in terms of one-electron (f_i) and two-electron operators (g_{ij}), one has

$$\mathbf{H} = \sum_i f_i + \sum_{i<j} g_{ij} \tag{5.6.1}$$

and the expectation value for an antisymmetric wave-function $\Psi(1,2...N)$ is

$$<\mathbf{H}> = \int dx_1 [f_1 \gamma(1,1')]_{1=1'} + \int dx_1 dx_2 [g_{12} \Gamma(12,1'2')]_{1'=1,2'=2} \tag{5.6.2}$$

where

$$\gamma(1,1') = N \int \Psi(12...N) \Psi^*(1'2'...N)\, dx_2...dx_N \tag{5.6.3}$$

is the first-order density matrix and

$$\Gamma(12,1'2') = \frac{1}{2} N(N-1) \int \Psi(12...N) \Psi^*(1'2'...N)\, dx_3...dx_N \tag{5.6.4}$$

is the second-order density matrix. In this formula x_i represents the space and spin coordinates of the ith electron.

The eigenfunctions of γ are called natural spin orbitals u_i and are defined by

$$\int dx_{1'} \gamma(1,1')\, u_i(1') = n_i\, u_i(1) \tag{5.6.5}$$

The density matrix can be expanded in terms of the natural orbitals by

$$\gamma(1,1') = \sum_i n_i\, u_i(1)\, u_i^*(1')$$

The diagonal part of γ is the one-electron probability distribution and one has

$$\rho(1) = \gamma(1,1) = \sum_i n_i\, u_i(1)\, u_i^*(1)$$

where $\rho(1)$ is the charge density for electron 1. Now it is possible to write the first term of $<\mathbf{H}>$ in terms of u_i

$$\int dx_1 \, [f_1 \, \gamma(1,1')]_{1=1'} = \sum_i n_i < u_i | f_1 | u_i >$$

The Coulombic two-electron operator is $g_{12} = r_{12}^{-1}$ and the second term of $<\mathbf{H}>$ is

$$\int dx_1 \, dx_2 \, [g_{12} \, \Gamma(12,1'2')]_{1=1',\,2=2'} = \int dx_1 \, dx_2 \frac{\Gamma(12,12)}{r_{12}}$$

Then the total energy may be written as

$$<\mathbf{H}> = \sum_i n_i < u_i | f_1 | u_i > + E_C + E_X \qquad (5.6.6)$$

where E_C, the Coulombic energy, is

$$E_C = \frac{1}{2} \int \frac{dx_1 \, dx_2 \, \rho(1)\rho(2)}{r_{12}} = \frac{1}{2} \int dx_1 \, \rho(1) \, U_C(1)$$

where U_C is the one-electron Coulomb potential

$$U_C(1) = \int \frac{\rho(2) \, dx_2}{r_{12}}$$

The exchange energy E_X is defined by

$$E_X = \frac{1}{2} \int \rho(1) \, U_x(1) \, dx_1 \qquad (5.6.7)$$

where $U_X(1) = \int \frac{\rho_x(1,2) dx_2}{r_{12}}$ is the exchange-correlation potential.

In the X_α method, this potential is defined as: $U_{X\alpha} = -(9/2)\alpha(3\rho/8\pi)^{1/3}$ where α is an adjustable parameter. This form is obtained after different approximations.

Finally, the energy $E_{X\alpha}$ takes the form

$$E_{X\alpha} = <\mathbf{H}> = \sum_i n_i < u_i(1) | f_i(1) | u_i(1) > + \frac{1}{2} \int \rho(1) U_C(1) dx_1 + \frac{1}{2} \int \rho(1) U_{X\alpha}(1) dx_1$$

The variational principle is applied to this form of $E_{X\alpha}$. The spin orbitals $u_k(1)$ are solutions of the one-electron Schrödinger equations

$$(f_1 + V_C + V_{X\alpha}) u_k = \varepsilon_k \, u_k \qquad (5.6.8)$$

and it has been shown that the one-electron energy eigenvalues ε_k are also equal to derivatives of the total energy with respect to the spin-orbital occupation numbers n_k:

$$\varepsilon_k = \partial \langle E_{X\alpha} \rangle / \partial n_k$$

Slater (1972) has proved that the total energy expression satisfies the virial and Hellmann-Feymann theorems. Matter is divided into clusters of atoms and in these local regions the potential is spherically or volume averaged and these approximations assure a rapidly convergent resolution.

The eigenvalues thus obtained can not be compared directly with the photoelectron spectra because Koopmans' theorem is not satisfied by the $X\alpha$ model.

The $X\alpha$-method was applied to cluster compounds particularly because other approximate methods did not seem to give interesting results. For example in a calculation on $Mo_2Cl_8^{4-}$ (Norman and Kolari 1975; Cotton 1978) it was possible to show that the Mo - Mo bond is a quadruple bond as was predicted by symmetry-based arguments. It was also possible to explain the bent structure of the complex $Rh_2Cl_2(CO)_4$ (Norman and Gmur 1977). The molecule

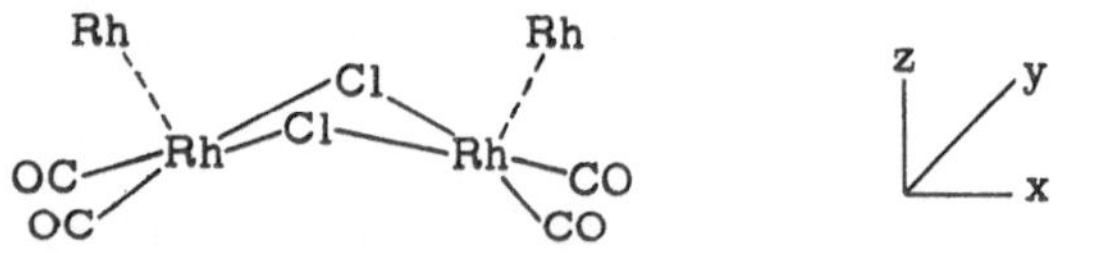

is bent along the Cl - Cl axis, the two-square-planar Rh $Cl_2(CO)_2$ units intersecting at an angle of 124°. The SCF - $X\alpha$ calculation showed that the lowering of symmetry converts a very weak Rh - Rh π interaction in one of the six Rh Cl bonding levels into a stronger one which is mainly σ in character.

In a recent calculation, the cluster compounds Re_3Cl_9, $Re_3Cl_{12}^{3-}$ and $Mo_6Cl_{14}^{2-}$ were studied (Cotton and Stanley 1978).

The calculation on Re_3Cl_9 confirms symmetry arguments on metal and ligand orbitals. In the two rhenium species, the metal atoms have a formal oxidation number of + 3 and contribute a total of twelve electrons to the formation of metal-metal bonds. In consequence there are double bonds between each pair of metal atoms. The corresponding irreducible representations of D_{3h} are a_1', e', a_2'', e''. To the first two correspond the Re - Re σ bonds and to the last two ones the Re - Re π bonds. It was found that, in general, each of the metal orbitals enters principally into two MO's, both of which are more or less metal-metal bonding in character, while the upper one is metal-ligand antibonding and the lower one metal-ligand bonding. In $Re_3Cl_{12}^{3-}$, the positions of the orbitals responsible for metal-metal bonding are only slightly affected because the interaction between the rhenium atoms and these terminal chloride ligands is weak.

In the cluster $Mo_6 Cl_{14}^{2-}$, each Mo atom is in the formal oxidation state + 2 and the cluster can be seen as built up with twelve Mo - Mo bonds of order unity along the edges of an octahedron. It was also found that the highest orbital is nearly Mo - Cl antibonding and a good agreement was found between the first experimental and theoretical UV spectrum band.

Due to the large number of electrons involved in the compounds containing metal-metal bonds there have been only few attempts to employ ab initio methods (Cotton and Stanley 1978; Benard and Veillard 1977; Garner *et al.* 1976; Serafini *et al.* 1978; Bénard *et al.* 1978). Recently calculations have been made for binuclear complexes of Cr and Mo metals in order to see if it was possible of describing the metal-metal bonding in terms of a quadruple bond. Such a description is based on the following sequence of metal-metal orbitals $\sigma < \pi < \delta < \delta^* < \pi^* < \delta^*$ with a ground-state configuration $(\sigma)^2 (\pi)^4 (\delta)^2$. An ab initio calculation at the SCF level with a double-zeta basis set for the valence shells gives a ground state for the molybdenum systems $Mo_2(O_2CH)_4$ and $[Mo\,Cl_8]^{4-}$ with $(\sigma)^2 (\pi)^4 (\delta)^2$ configuration in agreement with the X_α results [36]. The theoretical Mo - Mo bond length of 2.10 Å for $Mo_2(O_2CH)_4$ is in excellent agreement with the experimental one of 2.091 Å. In the corresponding chromium compounds at the SCF level the ground state is found in a non-bonding configuration $(\sigma)^2 (\delta)^2 (\delta^*)^2 (\sigma^*)^2$ as found previously by Garner *et al.* (1976), but after configuration interaction the configuration $(\sigma)^2 (\pi)^4 (\delta)^2$ becomes the leading term in the ground state wave function of these compounds. The great Cr - Cr bond lengths may be related to the large weight of the configurations other than the quadruple bonding in the configuration interaction expansion and the fact that $[Cr_2\,Cl_8]^{4-}$ does not exist may be related to the small energy gap between the bonding and non-bonding configurations at the minimum of the potential energy curve.

Thus for the elements of the first transition series it is necessary to include configuration interaction in the ab initio calculation for getting results in agreement with experiments. Very likely it would be the same if one had to compute electronic excited states. So far one ab initio calculation of the spectra of this kind of compounds has been published (Bénard 1978). It seems desirable that the future computations take account of the relativistic corrections (Pyykkö 1978).

5.7. Reliability of Computation and Choice of the Orbital Basis.

5.7.1. The ZDO assumptions.-

The ZDO (zero-differential overlap) approximation has been recalled several times in the present text. It consists in assuming that the product of the wave functions $\chi_\mu(\underset{\rightarrow}{n})$, $\chi_\nu(\underset{\rightarrow}{n})$ representing two different atomic orbitals in configuration space is everywhere zero

$$\chi_\mu(\underset{\rightarrow}{n})\chi_\nu(\underset{\rightarrow}{n}) = 0 \qquad \text{for } \mu \neq \nu \tag{5.7.1}$$

(zero-differential-overlap assumption)

Jug's (1969) excellent summary should be consulted in connection with the justification of that approximation; for recent work see e.g. Archirel and Barbier (1978). Here we just remind the reader of the fact that the ZDO assumption is a typical case of a possibly contradictory assumption. If the basis orbitals really satisfy the zero-differential-overlap condition, the core potential energy matrix is diagonal, and the mixing of atomic orbitals should depend on the core potential only through the diagonal elements of the effective (SCF) Hamiltonian. To assume, as is the case with most methods of the CNDO family, that the core potential matrix has non-vanishing off-diagonal elements, indeed related to overlap, implies that the theoretical consistency of the model is ensured only if there are MVAO's such that the corresponding two-electron integrals satisfy the ZDO condition, but the overlap integrals do not vanish. This is not an easy condition to meet, especially if one wants to satisfy it with transferable orbitals. However, in view of the many arguments presented in the literature, one should expect that such orbitals do exist; in fact, they could be defined for standard environments as, say, linear combinations of enough Gaussian orbitals to provide, with their exponents, a sufficient number of free parameters. It does not seem that the actual work has been done; until it is, the PPP-CNDO family can be considered a family of physical models only with some reservations.

5.7.2. Ab initio methods as semiempirical methods.

The well-known ab initio methods have been somewhat uncautiously presented as 'foolproof', 'non-empirical', 'accurate' methods. We shall accept here the practical point that they form a separate class of methods, but we point out that, in fact, a reasonable definition of any of the current ab-initio methods is first an SCF method which computes everything accurately *from a previously assigned basis.*

Thus, the choice of the basis is a crucial point, and is left to the intuition of the individual researcher.

That ab initio computations are especially sensitive to the choice of the basis was illustrated extensively by David (1970, 1971) and by David and Mély (1970), who discussed in detail the choice of the geometry, of the orbital exponents, of the size of the basis. Table 5.1.a,b,c illustrate the differences in sensivity of orbital energies, net charges, and total energy to changes in orbital exponents.

Table 5.1.a Orbital Exponents for STO-3G Calculations

Atom	type	(1)	(2)	(3)
Carbon	1S	5.67000	5.95300	6.23700
Oxygen	1S	7.66000	8.04300	8.42600
Hydrogen	1S	1.24000	1.30000	1.48800

Table 5.1.b Orbital Energies for Biacetyl with the Exponents of Table 5.1.a

(1)	(2)	(3)
-20.3081511	-19.9066015	-19.3617951
-20.3081505	-19.9065428	-19.3616180
-11.1360910	-10.8788454	-10.5332075
-11.1358897	-10.8750628	-10.5235706
-11.0497347	-10.8501465	-10.5083856
-11.0497252	-10.8500697	-10.5074668
- 1.3699395	- 1.2919830	- 1.1701495
- 1.3482659	- 1.2706463	- 1.1497974
- .9729954	- .9138180	- .8095574
- .9576413	- .9064087	- .8077619
- .8256126	- .7614208	- .6552934
- .6468928	- .5856624	- .4842581
- .6282313	- .5681072	- .4691997
- .5882677	- .5324522	- .4416572
- .5845117	- .5249569	- .4299938
- .5526232	- .5102782	- .4236167
- .5285768	- .4818671	- .3875010
- .5162786	- .4588353	- .3682802
- .4926075	- .4331989	- .3333302
- .4663673	- .3992762	- .2951319
- .4098690	- .3336979	- .2250817
- .3982520	- .3110699	- .1935621
- .3051815	- .2301293	- .1225907
.2048706	.3033062	.4318374
.3849331	.4754555	.5975152
.5872713	.6833596	.8322242
.6621180	.7543280	.9091886
.6689464	.7574361	.9252877
.7122811	.7972503	.9787132
.7132755	.7991323	.9861403
.7205800	.8058947	.9922742
.7323291	.8211339	1.0182469
.7414902	.8338964	1.0552591
.7447919	.8617130	1.0596207
.9496720	1.0727211	1.2377607

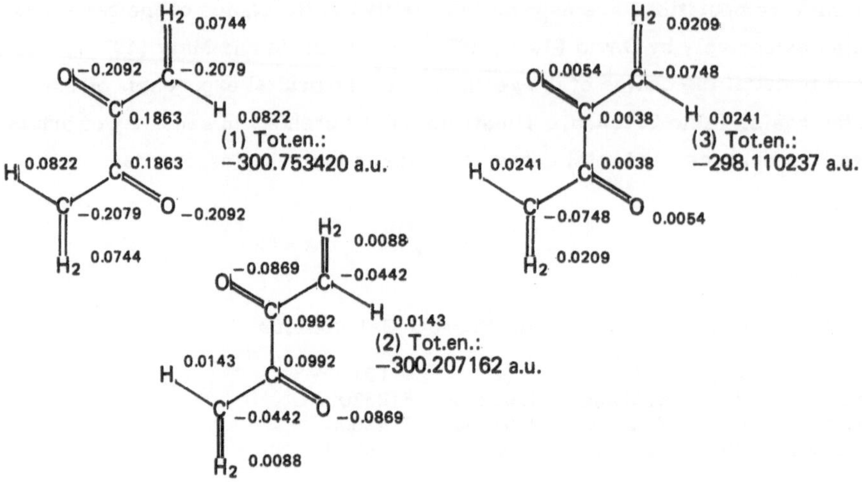

Table 5.1.c STO 3G Net Charges and Total Energies of Biacetyl With Different Sets of Slater Orbital (cfr. Table 5.1.a)

Criteria of convergency could be invoked to show that certain ab initio bases are actually non-empirical. Minimization of the energy with respect to the orbital exponents could be a good way to eliminate that source of ambiguity, extension of the basis until inclusion of more terms does not change the energy (or the populations) could eliminate the dependence on the size of the basis; finally, optimization of the geometry by the variational theorem should suppress the dependence of the basis on geometry.

Unfortunately, all the above optimizations are practically unfeasible, at least for comparatively large molecules. Moreover, they could be valid only for ground states and under two assumptions:

1) the class of basis functions over which convergency has been obtained is sufficiently complete - i.e., it is not possible that inclusion of the continuum, as advocated by Löwdin from time to time, would change the situation dramatically;

2) correlation effects are not important; in particular, they do not matter as regards the position of the energy minimum in configuration space.

It remains true that, especially in view of the inconsistencies in parametrization which affect semiempirical models, *ab initio* methods are often more reliable from the quantitative point of view (cf. sec. 4.9); of course, the price to pay is in physical interpretation, as is shown by difficulties encountered with a posteriori localization (Millié *et al.* 1975).

5.8. A comment on the Use of Group Theory in Calculations on Molecules and Aggregates of Atoms.

The real role of group theory in quantum chemistry has long been a subject of controversy (cf. "die Gruppentheoriepest" of H. Weyl). In fact, polyatomic molecules having more than one symmetry axis are not so common, and anyway the application of group theory to molecular systems is a comparatively simple matter; the only problem is to apply representation theory in order to construct bases for irreducible representations of known groups, point (and space) groups for geometry problems, permutation groups for spin problems. An excellent analysis of the latter has been provided by Matsen (1964). As regards geometry, group theory yields the space symmetry properties of the states associated with the various energy levels to be found without computations, in other words the spectroscopic label of those states. Hopefully, adaptation to symmetry also facilitates the determination of approximate wave functions, but the actual advantages depend on the problem at hand.

In the case of fixed-nuclei molecular problems which satisfy the Borh-Oppenheimer approximation without any special constraints, the study of space symmetries of the electronic system is especially easy, for the groups to be considered are just the point groups of the nuclear frames, either finite groups, or (for linear molecules) infinite but compact groups. In the latter case, contrary to a rather common opinion, it is not really necessary to limit one's considerations to an *ad hoc* finite subgroup of the group $D_{\infty h}$ or $C_{\infty v}$ in order to study the symmetry properties of those molecules; in fact, the technique of representation theory applies to any compact group, provided the invariant measure concept (cf. J. P. Serre 1977) is used. Complications arise only with non-rigid molecules, whose symmetry properties are studied only within the Schrödinger supergroup (Longuet-Higgins 1963, Altmann 1967); these problems are extremely interesting in molecular theory, but a discussion of them is outside the scope of the present article, and the reader should consult other reviews (J. Serre 1974). The molecular one-electron states $|W_1>$, $|W_2>, \ldots, |W_m>$ of molecular orbital theory are written in terms of a basis of m atomic orbitals spanning an m-dimensional vector space V_m eigenstates of an effective one-electron Hamiltonian (e.g. the Hückel Hamiltonian), and therefore they form another (orthonormal) basis in V_m, connected by a linear transformation to the atomic orbital basis. As the effective Hamiltonian is invariant under symmetry operations mapping the molecule onto itself, the molecular orbitals must transform in a well specified way under the operations of the point group G to which the molecule belongs; therefore, group theory can be used to construct sets α of symmetry-adapted combinations $|w_j>$ of tne atomic orbitals, such that the $|W_j>$'s are linear combinations only of the functions belonging to a given set.

The group G permutes the atomic orbitals $|\mu>$, and thus applies G to itself and defines a linear representation of G. Consider now the unitary irreducible representations Γ_α of G; it is possible to find a basis $|v_1>,\ldots,|v_m>$ in V_m which transforms as the basis vectors of one of the representations Γ_α. If $|v_j>$ and $|v_k>$ correspond to basis vectors of different representations, the following equations are satisfied:

$$<v_j|v_k> = 0 \qquad (5.8.1)$$

$$<v_j|\hat{H}^{eff}|v_k> = 0; \qquad (5.8.2)$$

These conditions hold also if $|v_j>$ and $|v_k>$ correspond to orthonormal basis vectors of the same representation (of dimension $n_\alpha>1$). Equations (5.8.1) and (5.8.2) are valid in particular for the basis $|w_1>, |w_2>,\ldots,|w_m>$ formed by the molecular orbitals, but hold for any orthogonal basis $|v_2>, |v_2>,\ldots,|v_m>$ of V_m, for they are just a consequence of the fact that, because of its invariance properties, H^{eff} commutes with the elements of G. They imply that the $|v_j>$ vectors belonging to the same representation provide a decomposition of V_m in a direct sum of invariant subspaces V_α. Equation (5.8.2) means that the matrix $\mathbf{H}^{eff}$ representing the operator $\hat{H}^{eff}$ over the symmetry-adapted basis is block-diagonal, each block $\mathbf{H}_\alpha^{eff}$ having eigenvalues of order n_α equal to the dimension of T_α, with eigenvector corresponding for of them to the basis vectors $|v_j> \equiv |v_{\alpha\omega}>$ ($\omega = 1,\ldots,n_\alpha$). The subscripts α and ω characterize the possible symmetry properties of molecular wave functions: they correspond to the azimuthal and magnetic quantum numbers of the hydrogen atom, respectively, the principal quantum number being replaced by a number ordering levels with the same symmetry (Roothaan 1951, footnote 31, p. 56).

The practical application of representation theory to the molecular orbital methods is carried out as follows. From a set of m atomic orbitals $|\mu\rangle$ symmetry orbitals $|\xi\rangle$ (also called group orbitals in ligand-field theory) are constructed by the relation

$$|\xi_j\rangle = \sum_r |\mu_r\rangle \lambda_{rj} \tag{5.8.3}$$

where the λ's are chosen so that the $|\xi_j\rangle$'s transform like the $|v_j\rangle$'s under the operations of the molecular symmetry group. The molecular orbitals are then given by linear combinations of symmetry orbitals

$$|w_k\rangle = \sum_j |\xi_j\rangle c_{jk} \tag{5.8.4}$$

Symmetry properties of the molecular orbitals are explicitly determined by the $|\xi_j\rangle$'s entering (5.8.4) for a given k.

If the symmetry properties of the orbitals entering a given electron configuration are known, the representation to which the total wave function belongs is determined by studying the transformation properties of the product of the occupied orbitals $|w_k\rangle$. It can be proved that a "closed-shell" system belongs to the totally symmetric representation, so that it is enough to consider the orbitals $|k\rangle$ of shells not completely filled.

The first stage of the calculation consists in determining the irreducible representations Γ_α contained in the reducible representation Γ_{red} generated by the primitive orbital basis. As long as point groups are under consideration, a well-known formula can be used to determine the number m_α of times (including the value 0) Γ_α is present in Γ_{red}:

$$m_\alpha = \frac{1}{g} \sum_s \varkappa(s)\varkappa_\alpha^*(s) \tag{5.8.5}$$

where $\varkappa(s)$ and $\varkappa_\alpha^*(s)$ are the characters of the operation s in Γ_{red} and Γ_α, respectively, and the summation runs over the g operations of the group. Note that not all the irreducible representations of G may appear in a given basis. For instance, out of the eight representations of the group D_{2h} of ethylene, only six are found when one considers just s and p orbitals on the carbon atoms, and s orbitals on the hydrogen atoms; a valence orbital basis of the EHT or CNDO type provides for the ground state the configuration

$$1a_g^2\ 1a_u^2\ 1b_{3u}^2\ 2a_g^2\ 1b_{2g}^2\ 1b_{2u}^2$$

plus the sequence of virtual orbitals

$1b_{3g}\ 1b_{3u}\ 3a_g\ 2a_u\ 3a_u\ 2b_{2g}$

The second step is the construction of basis orbitals $|\xi_j>$ from the atomic orbitals $|\mu_r>$. To that purpose, again if point groups are under consideration, the technique of projectors can be used (Altmann 1962):

$$|\mu>_{proj.,\alpha} = \frac{n_\alpha}{g} \sum_s |R_s(\mu)> \varkappa_\alpha^*(s) \qquad (5.8.6)$$

where $|R_s(\mu)>$ denotes the result of the application of the group operation s to $|\mu>$. Knowing the projection of $|\mu>$ in Γ_α, a linear combination of atomic orbitals entirely contained in that irreducible representation can be formed by inspection. If the dimension of Γ_α is larger than 1, different basis vectors can be obtained by replacing the character $\varkappa_\alpha$ by suitable (diagonal) elements of the corresponding unitary matrix U_α. (The character $\varkappa_\alpha$ and the diagonal elements of U_α correspond tó projection operators, not the off-diagonal elements of U_α.).

Similar formulas are available for compact groups corresponding to linear molecules, integration over angular parameters defining the operations of the groups $D_{\infty h}$ and $C_{\infty v}$ being necessary instead of the summation over s (J.P. Serre, loc. cit. p. 39).

The decomposition of the space V_m generated by an orbital basis into subspaces V_α is also useful for the calculation of observables. For instance, let a Cartesian reference system 0 x y z be given, the origin 0 being invariant with respect to the operations s of the group; in problems involving the transition moment, matrix elements of the form

$$< w_k | \hat{O} | w_l >$$

are needed, where $\hat{O}$ is either multiplication by or derivation with respect to a coordinate. If $|w_k>$ and $|w_l>$ are elements of V_m belonging to the irreducible representations Γ_α and Γ_α', and if T denotes the representation generated by the given coordinate, the above matrix element will be different from zero only if the tensor product (i.e. multiplication element by element)

$$\Gamma_\alpha \otimes T \otimes \Gamma_\alpha'$$

contains the unit representation.

The above remarks refer to point groups. The groups of solids (including polymers) are space groups whose elements include the translation operations, along one, two or three axes. They are infinite - non-compact - groups. The best procedure to construct symmetry orbitals in such cases consists in building Bloch orbitals (adapted to translational symmetry) whose local factor is a symmetry orbital of the unit cell. However, this may not be sufficient to give a basis for the irreducible representations of the entire space group, because there may be operations that depend on the particular type of lattice but are not just translations or point operations.

It is possible to use the entire symmetry of such systems by contructing explicitly their irreducible representations. Tables of those representations are available (e.g. Kovalev, 1965).

Acknowledgements. This article was complete when one of the authors (GDR) was visiting the Lehrstuhl für Theoretische Chemie of the University of Erlangen-Nürnberg as Richard-Merton Gastprofessor; the invitation by the Deutsche Forschungsgemeinschaft is hereby gratefully acknowledged.

1930-1949

Coulson, C.A., Longuet-Higgins, H.C.: The electronic structure of conjugated systems. I. General theory. Proc. Roy. Soc. A191, 39 (1947).

Coulson, C.A., Longuet-Higgins, H.C.: The electronic structure of conjugated systems. II. Unsaturated hydrocarbons and their hetero derivatives. Proc. Roy. Soc. A192, 16 (1947).

Coulson, C.A., Longuet-Higgins, H.C.: The electronic structure of conjugated systems. III. Bond orders in unsaturated molecules. Proc. Roy. Soc. A193, 447 (1948).

Coulson, C.A., Longuet-Higgins, H.C.: The electronic structure of conjugated systems. IV. Force constants and interaction constants in unsaturated hydrocarbons. Proc. Roy. Soc. A193, 456 (1948).

Coulson, C.A., Longuet-Higgins, H.C.: The electronic structure of conjugated systems. V. The interaction of two conjugated systems. Proc. Roy. Soc. A195, 188 (1948).

Coulson, C.A., Longuet-Higgins, H.C.: Cited in "dictionnaire des grandeurs théoriques descriptives des molécules", Revue Scient. 1947, 929.

Coulson, C.A., Rushbrooke, G.S.: Note on the method of molecular orbitals. Proc. Camb. Phil. Soc. 36, 193 (1940).

Goeppert-Mayer, M., Sklar, A.L.: Calculation of the lower excited states of benzene. J. Chem. Phys. 6, 645 (1938).

Heath, D.F., Linnett, J.W.: Molecular force fields. I. Structure of the water molecule. Trans. Far. Soc. 44, 556 (1948).

Heath, D.F., Linnett, J.W.: Molecular force fields. II. Force field of the tetrahalides of the Group IV elements. Trans. Far. Soc. 44, 561 (1948).

Herring, C: A new method for calculating wave functions in crystals. Phys. Rev. 57, 1169 (1940).

Hückel, E: Zur Quantentheorie der Doppelbindung. Z. Phys. 60, 423 (1930).

Klevens, H. B., Platt, J. R.: a.-Spectral resemblances of cata-condensed hydrocarbons. J. Chem. Phys. 17, 470 (1949).

b.-Spectral resemblances of cata-condensed hydrocarbons. J. Chem. Phys. 17, 472 (1949).

Mulliken, R.S.: a-Electronic structure of molecules. XI. Electroaffinity, molecular orbitals and dipole moments. J. Chem. Phys. 3, 573 (1935).

Mulliken, R.S.: b-Electronic structure of molecules. XII. Electroaffinity and molecular orbitals, polyatomic applications. J. Chem. Phys. 3, 586 (1935).

Mulliken, R.S.: The theory of molecular orbitals. J. Chem. Phys. 46, 497 (1949).

Pullman, A.: Comparaison des fréquences calculées et observées pour les deux plus longues bandes d'absorption de quelques hydrocarbures aromatiques polynucléaires. Compt. Rend. 229, 887 (1949).

Pullman, A., Berthier, G.: Structure électronique de l'azulène. Compt. Rend. 227, 677 (1948).

Pullman, A., Berthier, G.: Structure électronique, moment dipolaire et énergie de résonance de l'azulène. Compt. Rend. 229, 561 (1949).

Urey, H.C., Bradley, C.A., jr.: The vibrations of pentatomic tetrahedral molecules. Phys. Rev. 38, 1969 (1931).

Wheland, G., Mann, D.E.: The dipole moments of fulvene and azulene. J. Chem. Phys. 17, 264 (1949).

1950-1959

Anno, T.: Out-of-plane vibration of a conjugated hydrocarbon: transbutadiene. J. Chem. Phys. 28, 944 (1958).

Berthier, G., Pullman, B., Pontis, J.: Recherches théoriques sur les constantes de force et les fréquences de vibration du groupement carbonyle dans les molécules organiques conjuguées. J. Chimie Phys. 49, 367 (1952).

Berthier, G., Pullman, B.: Configuration geométrique et moment dipolaire des hydrocarbures conjugués. Bull. Soc. Chim. (France) D90 (1949).

Brown, R.D.: Molecular orbitals and organic reactions. Quart. Revs. 6, 63 (1952).

Chalvet, O., Daudel, R.: Etude théorique de la structure de la molécule N_2O_4. J. Chimie Phys. 49, 77 (1952).

Coulson, C.A., Duchesne, J.: La molecule N_2O_4 et un nouveau type de liaison chimique. Bull. Acad. Roy. Belg. Cl. Sci. 8, 522 (1957).

Daudel, A., Lefebvre, R., Moser, C.: Quantum chemistry. Methods and applications. New York: Interscience 1959.

Del Re, G.: Sulla struttura elettronica dell'ossido di benzonitrile. Rend. Acc. Naz. Lincei VIII 22, 491 (1957).

Del Re, G.: A simple MO-LCAO method for the calculation of charge distributions in saturated organic molecules. J. Chem. Soc. 1958, 4031.

Hubbard, J.: a-On the interaction of electrons in metals. Proc. Roy. Soc. 68, 441 (1955).

b-The dielectric theory of electronic interaction in solids. Ibid. 68, 976 (1955).

Hubbard, J.: The description of collective motions in terms of many-body perturbation theory. Proc. Roy. Soc. A240, 539 (1957).

Hubbard, J.: The description of collective motion in terms of many-body perturbation theory. II. The correlation energy of a free-electron gas. Proc. Roy. Soc. A243, 336 (1958).

Julg, A.: Etude de l'azulène par la méthode du champ moléculaire self-consistant. J. Chimie Phys. 52, 377 (1955).

Longuet-Higgins, H.C., Pople, J.A.: The electronic spectra of aromatic molecules. IV. Excited states of odd alternant hydrocarbons radicals and ions. Proc. Roy. Soc. (London) A68, 591 (1955).

Lykos, P.G., Parr, R.G.: On the pi-electron approximation and its possible refinement. J. Chem. Phys. 24, 1166 (1956).

Löwdin, P.O.: The nonorthogonality problem connected with the use of atomic wave functions in the theory of molecules and crystals. J. Chem. Phys. 18, 365 (1950).

Mataga, N., Nishimoto, K.: Electronic structure and spectra of nitrogen heterocycles. Z. Physik. Chem. (Frankfurt) 13, 140 (1957).

McLachlan, A.D.: Pairing of electronic states in alternant hydrocarbons. Mol. Phys. 2, 271 (1959).

Moffitt, W.E.: Aspects of hybridization. Proc. Roy. Soc. A202, 548 (1950).

Moffitt, W.E.: The electronic structure of the oxygen molecule. Proc. Roy. Soc. A210, 245 (1951).

Mulliken, R.S.: Electronic population analysis on LCAO-MO molecular wave functions. J. Chem. Phys. 23, 1833 (1955).

Orgel, L.E., Cottrell, T.L., Dick, W., Sutton, L.E.: The calculation of the electric dipole moments of some conjugated heterocyclic compounds. Trans. Far. Soc. 47, 113 (1951).

Pariser, R.: Electronic spectrum and structure of azulene. J. Chem. Phys. 25, 1112 (1956).

Pople, J.A.: Electron interaction in unsaturated hydrocarbons. Trans. Far. Soc. 49, 1375 (1953).

Pople, J.A., Nesbet, R.R.: Self-consistent orbitals for radicals. J. Chem. Phys. 22, 571 (1954).

Pritchard, H.O., Skinner, H.A.: The concept of electronegativity. Chem. Rev. 55, 745 (1955).

Pullman, B., Pullman, A.: Les théories électroniques de la chimie organique. Paris: Masson et Cie. 1952, pp. 665.

Roothaan, C.C.J.: New developments in molecular orbital theory. Rev. Mod. Phys. 23, 69 (1951).

Ruedenberg, K.: On the three- and four-center integrals in molecular quantum mechanics. J. Chem. Phys. 19, 1433 (1951).

Sandorfy, C.: LCAO MO (Linear Combination of Atomic Orbitals Molecular Orbitals) calculations on saturated hydrocarbons and their substituted derivatives. Can. J. Chem. 33, 1337 (1955).

Simpson, W.T.: Formal Hückel Theory. J. Chem. Phys. 28, 972 (1958).

Snyder, R.G., Hisatsune, I.C.: Infrared spectrum of dinitrogentetroxide. J. Mol. Spectr. 1, 139 (1957).

Smith, D.W., Hedberg, K.: Molecular structure of gaseous dinitrogen tetroxide. J. Chem. Phys. 25, 1282 (1956).

Walsh, A.D.: Electronic orbitals, shapes, and spectra of polyatomic molecules. Nature 170, 974 (1952).

Walsh, A.D.: The electronic orbitals, shapes, and spectra of polyatomic molecules, I-IX. J. Chem. Soc. 1953, 2260 (1953).

Wolfsberg, M., Helmholz, L.: The spectra and electronic structure of the tetrahedral ions MnO_4^-, CrO_4^{--}. J. Chem. Phys. 20, 837 (1952).

1960-1964

Aldous, J., Mills, I.M.: The calculation of force constants and normal co-ordinates. II. Methyl fluoride. Spectrochim. Acta 17, 719 (1962).

Anderson, R.W.: Localized magnetic states in metals. Phys. Rev. 124, 41 (1961).

Begun, G.M., Fletcher, W.H.: Infrared and Raman spectra of $N_2^{14}O_4$ and $N_2^{15}O_4$. J. Mol. Spectr. 4, 388 (1960).

Berthier, G.: Extension de la méthode du champ self-consistent à l'étude des molécules à couches électroniques incomplètes. Compt. Rend. 238, 91(1964).

Brown, R.D., Harcourt, R.D.: The electronic structures of A_2Y_4 molecules. Aust. J. Chem. 16, 737 (1963).

Crossman, L.D., Olsen, D.P., Duffey, G.H.: Bonding in the $Ta_6Cl_{12}^{2+}$ and $Mo_6Cl_8^{4+}$ structures. J. Chem. Phys. 38, 73 (1963).

De Heer, J.: Method of different orbitals for different spins and its application to alternant hydrocarbons. Rev. Mod. Phys. 35, 631 (1963).

Del Re, G.: a-On the non-orthogonality problem in the semi-empirical MO-LCAO method. Nuovo Cim. 17, 644 (1960).

b-An application of the semi-empirical MO-LCAO method to indoxazene and anthranil. Tetrahedron 10, 81 (1960).

Del Re, G.: a-Conjugation in unsaturated systems containing heteroatoms. Part I. The phenylisoxazoles. J. Chem. Soc. 1963, 3324

b-Hybridization and localization in the tight-binding approximation. Theoret. Chim. Acta (Berl.) 1, 188 (1963).

Del Re, G.: Theoretical procedures for the study of biochemical systems. In: Electronic aspects of biochemistry. B. Pullman (Ed.). Academic Press, New York (1964).

Del Re, G.: Hybridization and Localization in the Tight-Binding Approximation. Theoret. Chim. Acta 1, 188 (1963).

Del Re, G., Parr, R. G.: Toward an improved π-electron theory. Rev. Mod. Phys. 35, 604 (1963).

Del Re, G., Scarpati, R.: Sugli acidi pirroecarbonici. Nota XX. Considerazioni teoriche sulla reattività. Rend. Acc. Sci. Mat. Fis. (Napoli) 27, 512 (1960).

Dewar, M.J.S., Sabelli, N.L.: The split p-orbital method. III. Relationship to other M.O. treatments and application to benzene, butadiene, and naphtalene. Phys. Chem. 66, 2310 (1962).

Green, M., Linnett, J.W.: Molecules and ions containing an odd number of electrons. J. Chem. Soc. 1960, 4549 (1960).

Hisatsune, I.C., Devlin, J.P., Wada, Y.: Infrared spectra of some unstable isomers of N_2O_4 and N_2O_3. J. Chem. Phys. 33, 714 (1960).

Hinze, J., Jaffé, H.H.: Electronegativity. I. Orbital electronegativities of neutral atoms. J. Am. Chem. Soc. 84, 540 (1962).

Hinze, J., Jaffé, H.H.: a-Electronegativity. II. Bond and orbital electronegativities. J. Am. Chem. Soc. 85, 148 (1963).

b-Electronegativity. III. Orbital electronegativities and electron affinities of transition metals. Can. J. Chem. 41, 1315 (1963).

Hoffmann, R.: An extended Hückel theory. I. Hydrocarbons. J. Chem. Phys. 39, 397 (1963).

Jørgensen, C.K.: Absorption spectra and chemical bonding in complexes. Oxford: Pergamon Press 1962.

Le Goff, R., Serre, J.: Structure électronique de NO_2 et de N_2O_4. Theoret. Chim. Acta 1, 66 (1962).

Leman, G., Friedel, J.: Description of covalent bonds in diamond lattice structures by a simplified tight-binding approximation. J. Appl. Phys. 33, 281 (1962).

Longuet-Higgins, H.C.: The symmetry groups of non-rigid molecules. Molec. Phys. 6, 445 (1963).

Matsen, F.A.: Spin-free quantum chemistry. Adv. Quant. Chem. 1, 60 (1964).

McEwen, K.L.: Electronic structure of nitromethane and nitrogen dioxide. J. Chem. Phys. 32, 1801 (1960).

McWeeny, R.: The self-consistent generalization of Hückel theory. In: Molecular orbitals in chemistry, physics, and biology. (Löwdin, P.O., Pullman, B. eds.). New York: Acad. Press 1964.

Mills, I.M.: The calculation of accurate normal co-ordinates. Spectrochim. Acta 17, 719 (1961).

Moffat, J.B.: LCAO MO calculations on a series of nitriles and some related molecules by a method of self-consistent formal charges. Can. J. Chem. 42, 1323 (1964).

Mulliken, R.S.: Rydberg states of molecules. I-V. J. Am. Chem. Soc. 86, 3183 (1964).

Parr, P.: Quantum theory of molecular electronic structure. Reading, Mass.: Benjamin (1963).

Pauling, L.: The nature of the chemical bond. Ithaca: Cornell Univ. Press. (1960).

Platt, J. and co-workers: Systematics of the electronic spectra of conjugated molecules: a source book. New York: Wiley (1964).

Roothaan, C.C.J.: Self-consistent field theory for open shell of electronic systems. Rev. Mod. Phys. 32, 317 (1960).

Sandorfy, C.: Electronic spectra and quantum chemistry. Englewood Cliffs, N.J.: Prentice Hall (1964).

Scherer, J.R., Overend, J.: The application of the Urey-Bradley force field to the in-plane vibrations of benzene. Spectrochim. Acta 17, 719-(1961).

Streitwieser, Jr., A.: Molecular orbital study of ionization potentials of organic compounds, utilizing the ω-technique. J. Am. Chem. Soc. 82, 4123 (1960).

Streitwieser, A.: Molecular orbital theory for organic chemists. New York: Wiley (1961).

1965-1967

Altmann, S.L.: The symmetry of non-rigid molecules: the Schrödinger supergroup. Proc. Roy. Soc. A298, 184 (1967).

Berthier, G., Lemaire, H., Rassat, A., Veillard, A.: Interprétation structurale des spectres hyperfins de radicaux libres et méthodes de chimie quantique. Example des radicaux nitroxides. Theoret. Chim. Acta 3, 213 (1965).

Berthier, G., Millié, Ph., Veillard, A.: Recherches théoriques sur les complexes. I. Une méthode de calcul des orbitales moleculaires dans les complexes des métaux de transition. J. Chim. Phys. 62, 8 (1965).

Berthier, H., Pullman, A.: Sur le calcul des caracteristiques du squelette des molécules conjuguées. J. Chimie Phys. 62, 942 (1965).

Berthod, H., Giessner-Prettre, C., Pullman, A.: Sur les rôles respectifs des électrons σ et π dans les propriétés des dérivés halogénés des molécules conjuguées. Application à l'étude de l'uracil et du fluorouracil. Theoret. Chim. Acta 8, 212-22 (1967).

Blyholder, G., Coulson, C.A.: Molecular orbital models of chemisorption. Trans. Far. Soc. 63, 1782 (1967).

Cizek, J.: On the correlation problem in atomic and molecular problems. Calculation of wave function components in Ursell-type expansion terms using quantum field-theoretical methods. J. Chem. Phys. 45, 4256 (1966).

Cizek, J., Biczó, G., Ladik, J.: Some comments on the band structure calculation of linear chains in the semiempirical SCF-LCAO crystal orbital approximation. Theoret. Chim. Acta 8, 175 (1967).

Cusachs, L.C.: Semi-empirical molecular orbitals for general polyatomic molecules. II. One-electron model prediction of the H-O-H angle. J. Chem. Phys. 43, S 157 (1965).

De Brouckère, G.: Molecular orbital study of hydrated titanium ion. Bull. Soc. Chim. Belg. 76, 407 (1967).

Del Re, G.: On the choice and definition of atomic-orbital bases. I. General considerations; promotion and hybridization; electric dipole moments. Int. J. Quant. Chem. 1, 293 (1967).

Fischer-Hjalmars, I.: Connections between current π-electron theories. Theoret. Chim. Acta 4, 332 (1966).

Gélus, M., Vay, P.M., Berthier, G.: Structure électronique et stabilité thermique d'hétérocycles appartenant à la serie de l'imidazole et du thiazole. Theoret. Chim. Acta 9, 182 (1967).

Hinchliffe, A.: An omega-technique study of proton hyperfine coupling constants. Theoret. Chim. Acta 8, 300 (1967).

Jaffé, H.H., Beveridge, D. L., Orderin, M.: Understanding ultraviolet spectra of organic molecules. J. Chem. Educ. 44, 383 (1967).

Jørgensen, C.K., Horner, S.M., Hatfield, W.E., Tyree, Jr., S.Y.: Influence of Madelung (interatomic Coulomb) energy on Wolfsberg-Helmholz calculations. Int. J. Quant. Chem. 1, 191 (1967).

Kettle, S.F.A.: The Bonding within the $Mo_6Cl_8^{4+}$ and $Ta_6Cl_{12}^{2+}$ cations. Theoret. Chim. Acta 3, 211 (1965).

Koutecky, J.: a-Possible reinterpretation of Pariser-Parr-Pople theory. Chem. Phys. Letters 1, 249 (1967).

b-Some properties of semi-empirical Hamiltonians. J. Chem. Phys. 47, 1501 (1967).

Ladik, J.: a-Some developments in the semiempirical theory of molecular crystals. I. The Hückel approximation. Acta Phys. Hung. 18, 173 (1965).

b-Some developments in the semiempirical theory of molecular crystals. II. The Pariser-Parr-Pople approximation. Acta Phys. Hung. 18, 185 (1965).

Ladik, J.: A semiempirical method for the calculation of the excited states of molecules. Acta Phys. Hung. 23, 317 (1967).

Ladik, J., Biczó, G.: Investigation of the electronic structure of catalytically active solids. I. The change of the energy-band structure of infinite Ni crystals with the temperature. Acta Chim. Acad. Sci. Hung. 47, 263 (1966).

Longuet-Higgins, H.C.: Second quantization in the electronic theory of molecules. In: Quantum theory of atoms, molecules and the solid state. Löwdin, P.-O. (ed.). New York: Academic Press (1966) pp. 105-120.

Millié, Ph., Veillard, A.: Recherches theoriques sur les complexes. II. Complexes cyanés du fer, du cobalt, du nickel et composés analogues. J. Chimie Phys. 62, 20 (1965).

Momicchioli, F., Rastelli, A.: Benzo derivatives of the five-membered heterocycles. Theoretical treatment and UV spectra. J. Mol. Spectr. 22, 310 (1967).

Moore,Jr, E.B.: A population analysis of the bonding in N_2O_4, B_2Cl_4, C_2H_4 and C_3H_4. Theoret. Chim. Acta 7, 144 (1967).

Moseley, Wm. D. Jr., Ladik, J., Mårtensson, O.: On the propagation of errors in Hückel-Wheland molecular orbital calculations. Theoret. Chim. Acta 8, 18 (1967).

Newton, M.D., Boer, F.P., Lipscomb, W.N.: Molecular orbitals for organic systems parametrized from SCF model calculations. J. Am. Chem. Soc. 88, 2367 (1966).

Popkie, H.E., Moffat, J.B.: A quantum-theoretical study of some aromatic nitriles by the semi-empirical LCAO-SCF-MO method. Can. J. Chem. 43, 624 (1965).

Pople, J.A., Beveridge, D.L., Dobosh, P.A.: Approximate self-consistent molecular orbital theory. V. Intermediate neglect of differential overlap. J. Chem. Phys. 47, 2026 (1967).

Pople, J.A., Segal, G.A., Santry, D.P.: Approximate self-consistent molecular orbital theory. I. Invariant procedures. J. Chem. Phys. 43, S 129 (1965).

Pople, J.A., Segal, G.A.: Approximate self-consistent molecular orbital theory. II. Calculations with complete neglect of differential overlap. J. Chem. Phys. 43, S 138 (1965).

Pople, J.A., Segal, G.A.: Approximate self-consistent molecular orbital theory. III. CNDO results for AB_2 and AB_3 systems. J. Chem. Phys. 44, 3289 (1966).

Salem, L.: The molecular orbital theory of conjugated systems. New York: Benjamin (1966).

Veillard, A., Berthier, G.: π-Electron approach in pyridine and related compounds. Theoret. Chim. Acta 4, 347 (1966).

Veillard, A., Pullman, B.: a-Aspects de la structure électronique de la vitamine B_{12} et de ses analogues. J. Theoret. Biol. 8, 307 (1965).

b-Complements sur la structure électronique des ferroporphyrines biologiques. J. Theoret. Biol. 8, 317 (1965).

1968-1970

Abel, E.W., Stone, F.A.: The chemistry of transition-metal carbonyls: synthesis and reactivity. Quart. Rev. 24, 498 (1970).

Baird, M.C.: Metal-metal bonds in transition metal compounds. Prog. Inorg. Chem. 9, 1 (1968).

Berthier, G., Faucher, H., Gagnaire, D.: Un exemple de variation des constantes de couplage J_{C13-H} avec la charge atomique: l'ion dipropylcyclopropényle. Bull. Soc. Chim. (France) 1968, 1872 (1968).

Blyholder, G., Coulson, C.A.: Basis of extended Hückel formalism. Theoret. Chim. Acta 10, 316 (1968).

Buenker, R.J., Peyerimhoff, S.D.: Ab initio SCF calculation for azulene and naphtalene. Chem. Phys. Lett. 3, 37 (1969).

Cotton, F.A.: Strong homonuclear metal-metal bonds. Acc. Chem. Res. 2, 240 (1969).

David, D.J.: Détermination de bases de Gaussiennes optimales pour les molécules. II. Fonctions contractées pour la molécule d'hydrogène. Variation avec la distance interatomique. Theoret. Chim. Acta 19, 203 (1970).

David, D.J., Mely, B.: Détermination de bases de Gaussiennes optimales pour les molécules. I. Cas de l'hydrogène. Theoret. Chim. Acta 17, 145 (1970).

De Brouckère, G.: Molecular orbital studies of some transition metal complexes (bis π-2methylallyl metal). Theoret. Chim. Acta 19, 310 (1970).

Del Re, G., Ladik, J., Carpentieri, M.: On the effect of the inclusion of overlap in tight-binding band calculations of solids. Hung. Phys. Acta 24, 391 (1968).

Dewar, M.J.S., Haselbach, E.: Ground states of sigma-bonded molecules. IX. MINDO (Modified Intermediate Neglect of Differential Overlap)/2. J. Am. Chem. Soc. 92, 590 (1970).

Douady, J., Ellinger, Y., Rassat, A., Subra, R., Berthier, G.: Nitroxides XXXII. Analyse théoriques des effects de conformation sur la structure hyperfine des spectres R.P.E. Mol. Phys. 17, 217 (1969).

Ducastelle, F., Cyrot-Lackmann, F.: Moment developments and their application to the electronic charge distribution of bands. J. Phys. Chem. Solids 31, 1295 (1970).

Ellinger, Y., Rassat, A., Subra, R., Berthier, G.: Structure électronique et spectres de resonance paramagnetique électronique des radicaux vinyle et cyclopropyle. Theoret. Chim. Acta 10, 289 (1968).

Fischer-Hjalmars, I., Sundbom, M.: Semi-empirical parameters in π-electron systems. III. Heteroatomic molecules containing nitrogen. Acta Chim. Scand. 22, 607 (1968).

Grimley, T.B.: Overlap effects in the theory of adsorption using Anderson's Hamiltonian. J. Phys. C: Solid St. Phys. 3, 1934 (1970).

Harcourt, R.T.: Increased-valence theory of valence. J. Chem. Educ. 45, 779 (1968).

Heilbronner, E., Bock, H.: Das HMO-model und seine Anwendung/Tabellen berechneter und experimenteller Größen. Weinheim/Bergstr., Verlag Chemie (1970).

Hoffmann, R., Heilbronner, E., Gleiter, R.: Interaction of nonconjugated double bonds. J. Am. Chem. Soc. 92, 706 (1970).

Jaffé, H.H.: All-valence-electron semiempirical self-consistent field calculations. Accts. Chem. Res. 2, 136 (1969).

Jaffé, H.H.: All-valence-electron calculation of the electronic spectra of heterocyclic molecule. In: Quantum aspects heterocyclic compounds in chemistry and biochemistry. Jerusalem: Israel Ac. Sci. Hum. (1970).

Jug, K.: On the development of semi-empirical methods in the MO formalism. Theoret. Chim. Acta 14, 91 (1969).

Klopman, G., O'Leary, B.: All-valence-electrons SCF calculations. Fortschr. d. Chem. Forsch. 15, 445 (1970).

Koutecky, J., Michl, J., Becker, R.S., Erhart, C.E. Jr.: Note on the parameters for heteroatoms in Pariser-Parr-Pople (PPP) calculations. Theoret. Chim. Acta 19, 92 (1970).

Levine, R.D.: Quantum mechanics of molecular rate processes. Oxford: Clarendon Press (1969).

Magnasco, V., Musso, G.F.: The importance of intermolecular charge-transfer states. J. Chem. Phys. 48, 2657 (1968).

Moffat, J.B., Popkie, H.E.: Physical nature of chemical bond. II. Valence atomic orbital and energy partitioning studies of linear nitriles. Int. J. Quant. Chem. II, 565 (1968).

Momicchioli, F., Rastelli, A.: Theoretical studies on the ultraviolet spectra of five-membered heterocycles. π-Systems isoelectronic with condensed aromatic hydrocarbons. J. Chem. Soc. B 1970, 1353 (1970).

Nagy, J., Hencsei, P., Réfy, J.: Calculations on the sigma-bonding systems of organic silicon compounds. Acta Chim. (Budapest) 65, 51-57 (1970).

Newns, D.M.: Self-consistent model of hydrogen chemisorption. Phys. Rev. 178, 1123 (1969).

Packer, J.C., Avery, J.S., Ladik, J., Biczó, G.: The first excited triplet states of the nucleotide bases calculated with different semiempirical SCF schemes. Int. J. Quant. Chem. 3, 79 (1969).

Pulay, P.: Ab initio calculation of force constants and equilibrium geometries in polyatomic molecules. I. Theory. Mol. Phys. 17, 197 (1969).

Redmond, T.F., Wayland, B.B.: Comments on the planarity of dinitrogen tetroxide. J. Phys. Chem. 72, 3038 (1968).

Sandorfy, C.: Structure and spectra of σ-electron systems. Acta Phys. 27, 151 (1969).

Seprödi, L., Biczó, G., Ladik, J.: The effect of electric fields on the electronic structure of DNA. I. Calculation of the polarizability and of the permanent dipole moment for the nucleotide bases and base pairs. Int. J. Quant. Chem. 3, 621 (1969).

Silbermann, Z., Gershgorn, Z., Pauncz, R.: Application of the alternant molecular orbital method to non-alternant systems. Int. J. Quant. Chem. 2, (4), 453 (1968).

1971-1973

Asbrink, L., Fridh, G., Lindholm, E.: Interpretation of photoelectron spectra of hydrocarbons by use of a semiempirical calculation. J. Am. Chem. Soc. 94, 5501 (1972).

Baetzold, R.C.: Calculated properties of metal aggregates. I. Diatomic molecules. J. Chem. Phys. 55, 4355 (1971).

Baetzold, R.C.: Calculated properties of metal aggregates. III. Carbon substances. Surf. Sci. 36, 123 (1973).

Bailer, J.C., Emeleus, H.J., Hyholm, R., Trotman-Dickenson, A.F. (eds.): Comprehensive inorganic chemistry. Oxford: Pergamon Press 1973, vol. 2

Basso, J.H., Cabrol, D., Luft, R.: Ethylenic compounds. III. Determination of over-all electronic densities by CNDO (complete neglect of differential overlap)/2 and the sigma-π Pariser and Parr methods. Bull. Soc. Chim. (France) 4, 1388 (1973).

Bergmann, E.D., Pullman, D. (eds.): Aromaticity, pseudo-aromaticity, anti-aromaticity. Jerusalem: Isr. Acad. Sci. Hum. 1971.

Beveridge, D.L., Jano, I., Ladik, J.: INDO and MINDO/2 crystal orbital study of polyacetylene and polyglycine. J. Chem. Phys. 56, 4744 (1972).

Blyholder, G.: Quantum chemical treatment of adsorbed species. Mod. Aspects Electrochem. 1972, 1 (1972).

Clementi, E., Popkie, H.: Study of the structure of molecular complexes. I. Energy surface of a water molecule in the field of a lithium positive ion. J. Chem. Phys. 57, 3 (1972).

Cook, D.B.: Comments on a small Gaussian basis set in recent work in polymers. Chem. Phys. Lett. 11, 97 (1971).

Christoffersen, R.E.: Ab initio calculations on large molecules using molecular fragments. Benzene and naphtalene isomer characterizations and aromaticity considerations. J. Am. Chem. Soc. 93, 4104 (1971).

Daudel, R., Sandorfy, C.: Semiempirical wave-mechanical calculations on polyatomic molecules. New Haven, Conn.: Yale U. Press (1971).

David, D.-J.: Détermination de bases de Gaussiennes optimales pour les molécules. III. Remarques sur les bases atomiques de départ et application à la molécule de méthane. Theoret. Chim. Acta 23, 226 (1971).

De Brouckère, G., Trappeniers, N.J., Ten Seldam, C.A.: Calculation of the Fermi contact term in some transition metal complexes from unrestricted Hartree-Fock molecular orbitals. Chem. Phys. Lett. 21, 230 (1973).

Del Re, G., Momicchioli, F., Rastelli, A.: On the role of core repulsions in the prediction of the energy levels of π-conformers. Theoret. Chim. Acta 23, 316 (1972).

Dewar, M.J.S.: Aromatizität und pericyclische Reaktionen. Angew. Chem. 83, 859 (1971).

Diner, M.S.: Schéma de la méthode PCILO (Perturbative Configuration Interaction using totally Localized Orbitals). Colloq. Int. Cent. Nat. Rech. Sci. 195, 297 (1971).

Di Sipio, L., Tondello, E., De Michelis, G., Oleari, L.: Semi-empirical molecular orbital theory. The one-centre quantities for the elements of the first and second transition series. Chem. Phys. Lett. 11, 287 (1971).

Ellinger, Y., Rassat, A., Subra, R.: Structure électronique et couplages à longue distance dans un radical libre bicyclique. J. Chimie Phys. 68, 730 (1971).

Fliszar, S.: Charge distribution and chemical effects in saturated hydrocarbons. J. Am. Chem. Soc. 94, 1068 (1972).

Gayoso, J.: Contribution au problème de l'autocohérence dans le cadre de Hückel. III. Description d'une technique d'autocohérence non itérative. J. Chimie Phys. 68, 1096 (1971).

Gayoso, J.: Contribution au problème de l'autocohérence dans le cadre de Hückel. Description d'une méthode de Hückel autocohérente sans contrainte de spin. Compt. Rend. 274, 510 (1972).

Gilbert, M.M., Gundersen, G., Hedberg, R.: On a reinvestigation of the structure of dinitrogen tetrafluoride N_2F_4 by gaseous electron diffraction. J. Chem. Phys. 56, 1691 (1972).

Gropen, O., Seip, H.M.: Failure of the CNDO/2 method to predict the barriers and conformations in some conjugated systems. Chem. Phys. Lett. 11, 445 (1971).

Gupta, R., Majee, B.: a-Studies on organotin compounds using the Del Re method. IV. NMR spectra of organotin compounds. J. Organometal. Chem. 40, 97 (1972).

b-Studies on organotin compounds using the Del Re method. V. The chemical reactivity of organotin compounds. J. Organometal. Chem. 40, 107 (1972).

Herigonte, P.V.: Electron correlation in the seventies. Struct. and Bond. (Berl.) 12, 1 (1972).

Hoffmann, R.: Interaction of orbitals through space and through bonds. Accts. Chem. Res. 4, 1 (1971).

Hoffmann, R., Radom, L., Pople, J.A., Hehre, W.J., Salem, L.: Strong conformational consequences of hyperconjugation. J. Am. Chem. Soc. 94, 6221 (1972).

Huheey, J.E.: Inorganic chemistry: principles of structure and reactivity. New York, Harper and Row (1972).

Johnson, K.H.: Scattered wave theory of the chemical bond. Adv. Quant. Chem. 7, 143 (1973).

Kelkar, V.K., Bhalla, K.C., Khubchandani, P.G.: CNDO (complete neglect of differential overlap)/2 study of nitrogen oxides. J. Mol. Struct. 9, 383 (1971).

King, R.B.: Transition metal cluster compounds. Progr. Inorg. Chem. 15, 287 (1972).

Kutzelnigg, W., Del Re, G., Berthier, G.: σ and π electrons in theoretical organic chemistry. Fortschr. d. Chem. Forsch. 22, 1 (1971).

Ladik, J., Biczó, G.: a-Some developments in the semiempirical theories of molecular crystals. III. Different bands for different spins for π-electrons in the PPP approximation. Acta Chim. Ac. Sci. Hung. 67, 297 (1971).

b-Some developments in the semiempirical theories of molecular crystals. IV. σ-π (CNDO) bands. Acta Chim. Acad. Sci. Hung. 67, 397 (1971).

McClelland, B.W., Gundersen, G., Hedberg, K.: Reinvestigation of the structure of dinitrogentetroxide N_2O_4 by gaseous electron diffraction. J. Chem. Phys. 56, 4541 (1972).

Messmer, R.P., Watkins, G.D.: Molecular-orbital treatment for deep levels in semi-conductors. Substitutional nitrogen and lattice vacancy in diamond. Phys. Rev. B 7, 2568 (1973).

Moffat, J.B.: A semi-empirical quantum mechanical study of the dissociative chemisorption of hydrogen on a simulated boron surface. J. Coll. Interf. Sci. 44, 415 (1973).

Murrell, J.N.: The theory of the electronic spectra of organic molecules. London: Chapman and Hall (1971).

Murrell, J.N., Hatget, A.J.: Semi-empirical self-consistent-field molecular orbital theory of molecules. New York: Wiley (1972).

Nagy, J., Parkanyi, L.: Calculation of the oxygen-carbon σ-bond-system with modified Del Re constants. Acta Chim. (Budapest) 71, 159 (1972).

Penfold, C.R., Robinson, B.H.: Tricobalt carbon, an organometallic cluster. Accts. Chem. Res. 6, 73 (1973).

Perahia, D., Pullman, A.: Success of the PCILO method and failure of the CNDO/2 method for predicting conformations in some conjugated systems. Chem. Phys. Lett. 19, 73 (1973).

Port, G.N.J., Pullman, A.: Acetylcholine, gauche or trans? Standard ab initio self consistent field investigation. J. Am. Chem. Soc. 95, 4059 (1973).

Pozzoli, S.A., Rastelli, A., Tedeschi, M.: Hybridization and prediction of equilibrium geometries in alkanes, alkylamines, alkanols, and ethers. J. Chem. Soc. Far. Trans. 2, 69, 256 (1973).

Pulay, P., Török, F.: Calculation of molecular geometries and force constants from CNDO (Complete Neglect of Differential Overlap). Molec. Phys. 25, 1153 (1973).

Rössler, U., Treusch, J.: Semi-empirical band structure theory. Rep. Prog. Phys. 35, 883 (1972).

Salahub, D.R., Sandorfy, C.: CNDO, INDO and RCNDO-CI calculations on the electron spectra of saturated hydrocarbons. Theoret. Chim. Acta 20, 227 (1971).

Simonetta, M.: Qualitative and semiquantitative evaluation of reaction paths. Topics Curr. Chem. 42, 2 (1973).

Slater, J.C.: Hellman-Feynmann and Virial theorems in the X_α method. J. Chem. Phys. 57, 2389 (1972).

Suhai, S., Ladik, J.: CNDO/2 and MINDO/2 energy band structure of some homopolynucleotides. Int. J. Quant. Chem. 7, 547 (1973).

Trappeniers, N.J., De Brouckers, G., Ten Seldam, C.A.: Molecular orbital calculations on some copper complexes. Chem. Phys. Lett. 8, 327 (1971).

Vandorffy, M.T.: Neue Methode zur Bestimmung des δ^o-Parameters der quantenchemischen Del Re-Rechnungen. Acta Chim. (Budapest) 71, 139 (1972).

Yathindra, N., Rao, V.S.R.: Electronic charge distribution in mono-, di-, and polysaccharides. Carbonyl. Res. 25, 256 (1972).

1974-1976

Ahlrichs, R., Keil, F.: Structure and bonding in dinitrogen tetroxide (N_2O_4). J. Am. Chem. Soc. 96, 7615 (1974).

Allan, M., Heilbronner, E., Kloster-Jensen, E.: A photoelectronic spectroscopic investigation of benzologue tropones. J. Electr. Spectr. Relat. Phen. 6, 181 (1975).

Ames, D.L., Turner, D.W.: Photoelectron spectroscopic studies of dinitrogen tetroxide and dinitrogen pentoxide. Proc. Roy. Soc. (London) A348, 175 (1976).

Avery, J.: Creation and annihilation operators. Düsseldorf: McGraw Hill (1976).

Baetzold, R.C.: Application of molecular orbital theory to catalysis. Adv. Catal. 25, 1 (1976).

Baetzold, R.C., Mack, R.E.: Calculated properties of CuCl and Na clusters. Inorg. Chem. 14, 686 (1975).

Bingham, R.C., Dewar, M.J.S., Lo, D.H.: Ground states of molecules. XXVI. MINDO/3 calculations for hydrocarbons. J. Am. Chem. Soc. 97, 1294 (1975).

Batich, C., Heilbronner, E., Vogel, E: The ionization energies of bridged /4/ annulenes and of dicyclohepta(cd,gh)pentalene. Helv. Chim. Acta 57, 2288 (1974).

Biczó, G., Kertesz, M., Suhai, S.: Eine neue Methode zur Berechnung der Ladungverteilung in Polymeren unter Berücksichtigung der Endeffekte. Z. Chem. 15, 203 (1975).

Bieri, G., Heilbronner, E., Kobayashi, T., Goldstein, M.J., Leight, R.S., Lipton, M.S.: Dewar benzene and some of its derivatives. A photoelectron spectroscopic analysis. Helv. Chim. Acta 59, 2657 (1976).

Califano, S.: Vibrational states. London, Wiley 1976.

Cizek, J., Paldus, J., Hutac, H.: Correlation effects in the low-lying excited states of the PPP models of alternant hydrocarbons. I. Qualitative rules for the effect of limited configuration interaction. Int. J. Quant. Chem. 8, 951 (1974).

Cizek, J., Pellegatti, A., Paldus, J.: Correlation effects in the PPP model of alternant π-electronic systems: Two-point Padé approximant approach. Int. J. Quant. Chem. 9, 987 (1975).

Del Re., G.: The non-orthogonality problem and orthogonalization procedures. In: Quantum science methods and structure. Calais, J.L. et al. (eds.). Plenum Press, New York (1976), pp. 53.

Del Re, G., Lami, A.: Aspects of the quantum theory of chemical reactions. Bull. Soc. Chim. Belg. 85, 995 (1976).

Dewar, M.J.S.: Computing calculated reactions. Chem. In Britain 11, 97-(1975).

Doedens, R.J.: Structure and metal-metal interactions in copper (II) carboxylate compounds. Progr. Inorg. Chem. 21, 209 (1976).

Douady, J., Ellinger, Y., Subra, R.: Interaction de configuration perturbative en base d'orbitales localisées dans le cadre des approximations INDO. Bull. Soc. Chim. Belg. 85, (1976).

Durand, Ph., Barthelat, J.C.: Effective molecular hamiltonians, pseudopotentials and molecular applications. In: Localization and delocalization in Quantum Chemistry. O. Chalvet et al. (eds.). Dordrecht: Reidel 1976, vol. 2, pp. 91-125

Eady, C.R., Johnson, B.F.G., Lewis, J.: The chemistry of polynuclear compounds. Part XXVI. Products of the pyrolysis of dodeca-carbonyl-triangulo-triruthenium and triosmium. J. Chem. Soc. Dalton Trans. 2606 (1975).

Fliszar, S.: Charge distribution and chemical effects. IV. Criterion for selecting a theoretical method for the study of molecular properties involving charges. (and references therein). J. Am. Chem. Soc. 96, 4353 (1974).

Garner, D.C., Hillier, I.H., Guest, M.F., Green, J.C., Coleman, A.W.: The Nature of the Metal-Metal Interaction in Tetra-α-Carbonylatochromium (II) Systems. Chem. Phys. Lett. 41, 91 (1976).

Gayoso, J., Bouanani, H., Boucekkine, A.: Etudes en série pyronique. I. Application de la méthode de Hückel autocohérente et de la méthode semi-empirique de Pople à une série de pyrones et benzopyrones: comparaison de ces méthodes. Bull. Soc. Chim. (France) 3-4, 538 (1974).

Griffiths, R.L., MacLagan, R.G.A.R., Phillips, L.F.: Molecular orbital studies of dinitrogen tetroxide and related molecules. J. Chem. Phys. 3, 451 (1974).

Hay, P.J., Thibault, J.C., Hofmann, R.: Orbital interaction in metal dimer complexes. J. Am. Chem. Ioc. 97, 4884 (1975).

Heilbronner, E., Schmelzer, A.: A quantitative assessment of "through-space" and "through-bond" interactions. Application to semi-empirical SCF models. Helv. Chim. Acta 58, 936 (1975).

Howell, J.M., Van Wazer, J.R.: Electronic structures of dinitrogen tetroxide and diborane tetrafluoride and an analysis of their conformational stabilities. J. Am. Chem. Soc. 96, 7902 (1974).

Johnson, K.H.: Quantum Chemistry, in Ann. Rev. Phys. Chem. 26, 39 (1975).

Julg, A., Bernard, M., Bourg, M., Gillet, M., Gillet, E.: Adaptation of the molecular-orbital method to study the crystalline structure and shape of a monovalent metal. Application to the lithium. Phys. Rev. B 9, 3248 (1974).

Klinger, R.J., Butler, W., Curtis, M.D.: Synthesis, reactivity and molecular structure of cyclopentadienylmolbdenum dicarbonyl dimer. The molybdenummolybdenum triple bond. J. Am. Chem. Soc. 97, 3535 (1975).

Koutecky, J.: Quantum theory of surface phenomena. Progr. Surface and Membrane Sci. 11 (1976).

Koutecky, V.B., Koutecky, J.: General properties of the Hartree-Fock problem demonstrated on the frontier orbital model. Theoret. Chim. Acta 36, 149 (1975).

Ladik, J.: Semi-empirical energy band structures of periodic DNA and protein models. In: Electronic structure of polymers and molecular crystals. André, J.-M., Ladik, J. (eds.) New York: Plenum Press (1975), pp. 663-679.

Ladik, J.: Different LCAO band structure calculation methods for periodic polymers and molecular crystals. In: Electronic structure of polymers and molecular crystals. André, J.-M., Ladik, J. (eds.) New York: Plenum Press (1975), pp. 23-52.

Leroy, G., Sana, M.: Etude théorique de la cycloaddition dipolaire. Tetrahedron 31, 2091 (1975).

Levenson, R.A., Gray, H.B.: The electronic structure of compounds containing metal-metal bonds. Decacarbonyldimetal and related complexes. J. Am. Chem. Soc. 97, 6042 (1975).

Messmer, R.P., Knudson, S.K., Johnson, K.H., Diamond, J.B., Yang, R.P.: Molecular-orbital studies of transition and noble-metal clusters by the self-consistent-field-X_α scattered-wave method. Phys. Rev. 13, 1396 (1976).

Messmer, R.P., Tucker, C.W.Jr., Johnson, K.H.: Comparison of SCF-X_α and extended Hückel methods for metal clusters. Chem. Phys. Lett. 36, 423 (1975).

Miller, J.S., Epstein, A.J.: One dimensional inorganic complexes. Progr. in Inorg. Chem. 20, 1 (1976).

Millié, Ph., Levy, B., Berthier, G.: Localization and delocalization in orbital theories. In: Localization and delocalization in Quantum Chemistry. Chalvet, O. et al. (eds.) Dordrecht-Holland: Reidel (1975), vol. 1, p. 59.

Momicchioli, F., Corradini, G.R., Bruni, M.C., Baraldi, I.: Mechanism of the direct trans cis photoisomerization of stilbene. 2. Thermally activated intersystem crossing. J. Chem. Soc. Faraday Trans. 1975, 215 (1975).

Muda, Y., Hanawa, T.: Study of adsorption by the extended Hückel theory: CO in diamond (111) surface. Japanese J. of Appl. Phys. 13, 930 (1974).

Müller, H., Opitz, Ch.: Chemisorptionsprozesse als Vorstufe heterogener Reaktionsabläufe in quantenchemischer Sicht: Der Balloneffekt. Z. Phys. Chem. 257, 482 (1976).

Müller, H., Opitz, C.: Die beiden Möglichkeiten des Cluster-Modells bei der Untersuchung von Festkörpern. Z. Chem. 16, 200 (1976).

Norman, J.G.Jr., Kolari, H.J.: Electronic structure of octachlorodimolybdate (II) J. Am. Chem. Soc. 97, 33 (1975).

Pancir, J.: Equilibrium geometry and vibrational characteristics computations by semi-empirical methods. Coll. Czech. Comm. 40, 2726 (1975).

Pandey, K.C.: Realistic tight-binding model for chemisorption: H on Si and Ge (111). Phys. Rev. 14, 1557 (1976).

Richards, W.G., Raftery, J., Hinkley, R.K.: The calculation of spectroscopy constants. In: Theoretical chemistry. London: Chemical Soc. 1, 1 (1974).

Rouse, R.: The physical nature of the lone pair. I. Approaches to the problem. Theoret. Chim. Acta 41, 149 (1976).

Salem, L.: Surface crossing and surface touchings in photochemistry. J. Am. Chem. Soc. 96, 3486 (1974).

Sandorfy, C.: Far ultraviolet absorption spectra of organic molecules. Lone pairs and double bonds. In: Chemical spectroscopy and photochemistry in the vacuum ultraviolet. Sandorfy, D., Ausloos, P.T., Robin, M.B. (eds.) Boston, Mass.: Reidel (1974).

Pearson, R.: Symmetry rules for chemical reactions. Chem. in Britain 12, 146 (1976).

Politzer, P.: Some approximate energy relationship for molecules. J. Chem. Phys. 64, 4239 (1976).

Pullman, B. (ed.): Quantum mechanics of molecular conformations. Jerusalem: Israel Ac. Sci. Hum. (1976).

Sanderson, R.T.: Electronegativity equalization and partial charge. Educ. Chem. 11, 80 (1974).

Scott, J.M., Sutcliffe, B.T.: A configuration interaction study of phosphine using bonded functions. Theor. Chim. Acta 41, 141 (1976).

Serre, J.: Symmetry Groups of non rigid molecules. Adv. Quant. Chem. 8, 1 (1974).

Smeyers, Y.G., Brucena, A.M.: A method for direct derivation of unrestricted Hartree-Fock equations. Bull. Soc. Chim. Belg. 85, 1017 (1976).

Simon, A., Mattausch, H., Holzer, N.: Monochlorides of the Lanthanoids: Gd Cl and Tb Cl. Angew. Chem. (Int. Ed.) 15, 624 (1976).

Torrini, M., Zanazzi, E.: Correlation effects in small metal clusters. J. Phys. C 9, 63 (1976).

Tejeda, J., Shevchik, N.J.: Orbital nonorthogonality effects in band structure. Bond orbital model. Phys. Rev. B 13, 2548 (1976).

Wade, K.: Structural and bonding patterns in cluster chemistry. Adv. in Inorg. Chem. and Radiochem. 18, 1 (1976).

Wentworth, W.E., Wang,Kao, L., Becker, R.S.: Electron affinities of substituted aromatic compounds. J. Phys. Chem. 79, 1161 (1975).

1977-1980

Archirel, P., Barbier, C.: Réexamen,de l'approximation du recouvrement différentiel nul en théorie des constantes de couplage RMN. J. Chimie Phys. 75, 1003 (1978).

Adams, B.G., Paldus, J., Cizek, J.: Application of graphical methods of spin algebras to limited CI approaches. II. A simple open shell case. Int. J. Quant. Chem. 11, 849 (1977).

Bailey, W.I.Jr., Chisholm, M.H., Cotton, F.A., Murillo, C.A., Randel, L.A.: Reactions of metal-to-metal multiple bonds. 1 μ-Allene-bis(cyclopentadienyl) tetracarbonyl-dimolybdenum and -ditungsten compounds. Preparation, properties and structural characterization. J. Am. Chem. Soc. 100, 802 (1978).

Dantle, S., Ahlrichs, R.: Some limitations of the MINDO/3 method. Chem. Phys. Lett. 53, 148 (1978).

Barone, V., Del Re, G., Fliszar, S.: Bond energies and inductive effects. J. Chem. Soc. (Perkins II) in press (1979).

Barthelat, J.C., Durand, Ph.: Recent progress of pseudo-potentials methods in quantum chemistry. Gazz. Chim. It. 108, 225 (1978).

Basch, H., Bieri, G., Heilbronner, E., Jones, T.B.: The photoelectron spectrum of tetrafluorobutatriene. Helv. Chim. Acta 61, 46 (1978).

Bau, R., Carroll, W.E., Teller, R.G., Koetzle, T.F.: A quadruply hydrogen-bridged metal-metal bond. The neutron diffraction analysis of $H_8Re_2(P\ Et_2Ph)_4$. J. Am. Chem. Soc. 99, 3872 (1977).

Bénard, M., Veillard, A.: Nature of the metal-metal interaction in binuclear complexes of Cr and Mo. Nouv. J. Chim. 1, 97 (1977).

Bénard, M.: A theoretical study of the metal-metal interaction in binuclear complexes of transition groups 6 and 7. J. Am. Chem. Soc. 100, 2354 (1978).

Ben Lakdar, T., Suard, M., Taillandier, F., Berthier, G.: Vibrational properties of polyatomic molecules by quantum-chemical methods. I. Test-calculations on force constants and infrared intensities. Molec. Phys. 36, 509 (1978).

Bieri, G., Heilbronner, E., Jones, T.B., Kloster-Jensen, E., Maier, J.P: Some aspects of acetylene photoelectron spectroscopy. Phys. Scripta 16, 202 (1977).

Bischof, P., Eaton, P.E., Gleiter, R., Heilbronner, E., Jones, T.B., Musso, H., Schmelzer, A., Stober, R.: The electronic structure of Cubane (C_8H_8) as revealed by photoelectron spectroscopy. Helv. Chim. Acta 61, 547 (1978).

Bloch, M., Brogli, F., Heilbronner, E., Jones, T.B., Prinzbach, H., Schweikert, O.: Photoelectron spectra of unsaturated oxides. I. 1,4-Dioxin and related systems. Helv. Chim. Acta 61, 1388 (1978).

Catterick, J., Thornton, P.: Structures and Physical Properties of Polynuclear Carboxylates. Adv. Inorg. Chem. and Radiochem. 20, 291 (1977).

Chisholm, M.H., Cotton, F.A.: Chemistry of compounds containing metal-to-metal triple bonds between molybdenum and tungsten. Accts. Chem. Res. 11, 356 (1978).

Chisholm, M.H., Cotton, F.A., Extine, M.W., Kelly, R.L.: Reactions of metal-to-metal multiple bonds. 3. Addition of nitric oxide to hexakis (alkoxy) dimolybdenum compounds. Preparation and properties of bis (nitrosyl)-hexakis-(alkoxy)-dimolybdenum compounds and structural characterization of the isopropoxy derivative. J. Am. Chem. Soc. 100, 3354 (1978).

Cizek, J.: Coupled cluster many electron theory. In: Quantum theory of polymers. André, J.-M. et al. (eds.). Dordrecht: Reidel, 1978, p. 103.

Connolly, J.W.D.: The X_α method. In: Semiempirical methods of electronic structure calculations. Part A: Techniques. (Segal, G.A., ed.) New York: Plenum Press 1977, p. 105.

Cotton, F.A., Fanwick, Ph.E., Gage, L.D., Kalbacher, B., Martin, D.S.: Spectroscopic study of the $Tc_2Cl_8^{3-}$ ion. J. Am. Chem. Soc. 99, 5642 (1977).

Cotton, F.A.: Discovering and understanding multiple metal-to-metal bonds. Acc. Chem. Res. 11, 225 (1978).

Cotton, F.A., Stanley, G.G.: Ground state electronic structure of some metal atom cluster compounds. Chem. Phys. Lett. 58, 450 (1978).

Dalgaard, E., Jorgensen, P.: Optimization of orbitals for multiconfigurational reference states. J. Chem. Phys. 69, 3833 (1978).

De Brouckère, G.: Calculations of observables in metallic complexes by the molecular orbital theory. Int. J. Quant. Chem. 37, 203 (1978).

De Bruijn, S.: Comments on HAM/3, a semi-empirical MO theory. Chem. Phys. Lett. 52, 76 (1977).

De Bruijn, S.: Resonance integrals in semi-empirical MO theories. Chem. Phys. Lett. 54, 399 (1978).

Del Re, G., Julg, A., Lami, A.: On the theoretical foundations of simple quantum chemical treatments of clusters. J. Physique C38, 2 (1977).

Demitras, G.C., Muetterties, E.L.: Metal clusters in catalysis. 10. A new Fischer-Tropsch synthesis. J. Am. Chem. Soc. 99, 2796 (1977).

Douady, J., Ellinger, Y., Subra, R.: The PCILO method at the INDO level. Chem. Phys. Lett. 56, 38 (1978).

Douady, J., Barone, V., Ellinger, Y., Subra, R.: Perturbative configuration interaction using localized orbitals in the INDO hypothesis. I. Theory and applications to energetical problems. Int. J. Quan. Chem. 17, 211 (1980).

Ernst, R.D., Marks, T.J., Ibers, J.A.: Metal-metal bond cleavage reactions. The crystallization and solid-state structural characterization of cadmium tetracarbonyliron, Cd $Fe(CO)_4$. J. Am. Chem. Soc. 99, 2090 (1977).

Erwin, D.K., Geoffroy, G.L., Gray, H.B., Hammond, G.S., Solomon, E.I., Trogler, W.C., Zagars, A.A.: Production of hydrogen by ultraviolet irradiation of $Mo_2(SO_4)^{4-}$ in aqueous sulfuric acid. Electronic absorption spectrum of $K_3Mo_2(SO_4)_4$. 3,5 H_2O at 15 K. J. Am. Chem. Soc. 99, 3620 (1977).

Flanigan, M.C., Komornicki, A., McIver, J.W.Jr.: Ground state potential surface and thermochemistry. In: Modern Theoretical Chemistry. G.A. Segal (ed.) New York: Plenum Press 1971, vol. 8, pp. 1-45.

Graovac, A., Gutman, I., Tinajstic, N.: Topological approach to the chemistry of conjugated molecules. Lecture Notes in Chem. 4, 1 (1977).

Harcourt, R.D.: Configuration interaction and a new assignment for the second ionization potential of dinitrogen tetroxide. Chem. Phys. Lett. 61, 25 (1979).

Hermann, K., Bagus, P.S.: Binding and energy-level shifts of carbon monoxide adsorbed on nickel: model studies. Phys. Rev. B 16, 4195 (1977).

Itoh, H.: Molecular orbital calculations of adsorption of CO on Ni and Cu metal clusters. Japanese J. of Applied Phys. 16, 2125 (1977).

Kaga, H., Yoshida, K.: Orthogonality catastrophe due to local electron correlation. Progr. Theoret. Phys. 59, 34 (1978).

Kertesz, M., Kollar, J., Azman, A.: On Hartree-Fock orbital and total energies in extended systems. J. Chem. Phys. 69, 2937 (1978).

Kishner, S., Whitehead, M.A., Gopinathan, M.S.: Localized molecular orbitals for N_2O_2, N_2O_3 and N_2O_4. J. Am. Chem. Soc. 100, 1365 (1978).

Laine, R.M., Rinker, R.G., Ford, P.C.: Homogeneous catalysis by ruthenium carbonyl in alkaline solution: the water gas shift reaction. J. Am. Chem. Soc. 99, 252 (1977).

Lami, A., Del Re., G.: An incoherent exciton scattering model for the prediction of dimer bandshapes. Chem. Phys. 28, 155 (1978).

Masuda, K.: Changes in density of states caused by chemisorption: monolayer of of atoms on a model transition metal. Z. Naturforsch. 33a, 66 (1978).

Mathur, S.C., Mitra, S., Kumar, B., Singh, D.C.: Two-parameter ω-technique for MO calculations. Int. J. Quant. Chem. 11, 759 (1977).

Messmer, R.P., Salahub, D.R.: Chemisorption of oxygen atoms on aluminium (100): A molecular-orbital cluster study. Phys. Rev. 16, 3415 (1977).

Michaelson, H.A.: Relation between an atomic electronegativity scale and the work function. IBM J. Res. Dev. 22, 72 (1978).

Nicolas, G., Durand, Ph.: A new general methodology for deriving transferable atomic potentials in molecules. J. Chem. Phys., 70, 2020 (1979); Comm. "4th seminar on computational methods in quantum chemistry" Sept. 1978, Lund, Sweden.

Norman, Jr., J.G., Gmur, D.J.: Explanation for the band structure of $Rh_2Cl_2(CO)_4$. J. Am. Chem. Soc. 99, 1446 (1977).

Paldus, J., Adams, B.G., Cizek, J.: Application of graphical methods of spin algebras to limited CI approaches. I. Closed-shell case. Int. J. Quant. Chem. 11, 813 (1977).

Parr, R.G., Donnelly, R.A., Levy, M., Palke, W.E.: Electronegativity: the density functional viewpoint. J. Chem. Phys. 68, 3801 (1978).

Pauncz, R.: Branching diagram and serber-type spin functions. Algorithms for their construction and special properties. Int. J. Quant. Chem. 12, 369 (1977).

Pellegatti, A.: Simple extended STO basis sets. Helium. Int. J. Quant. Chem. 12, 545 (1977).

Pouchan, C.: Thèse de doctorat d'Etat. Université de PAU, 21 Mars 1978.

Pouchan, C., Dargelos, A., Chaillet, M.: Etude comparée des différents facteurs agissant sur la qualité des fonctions d'onde dans un calcul ab initio de champ de force. J. Chimie Phys. 75, 595 (1978).

Pyykkö, P.: Relativistic quantum chemistry. Adv. Quant. Chem. 11, (1978).

Reis, Jr., A.H., Hagley, V.S., Peterson, S.W.: Stabilization of one-dimensional conducting materials by carbonyl ligands. Crystal and molecular structure of $Ir(CO)_3Cl$. J. Am. Chem. Soc. 99, 4184 (1977).

Ruedenberg, K.: An approximate relation between orbital SCF energies and total SCF energy in molecules. J. Chem. Phys. 66, 375 (1977).

Salahub, D.R., Messmer, R.P.: Molecular-orbitals study of aluminium clusters Phys. Rev. 16, 2526 (1977).

Selsby, R.G., Grimison, A.: A semi-empirical theory for ionization potentials and electron affinities. II. Vertical and adiabatic values, benzenoid and nonbenzenoid aromatic hydrocarbons, and conjugated molecules with heteroatoms. Int. J. Quant. Chem. 12, 527 (1977).

Selsby, R.G., Machin, C., Hernandez, M.L.: A semi-empirical MO theory for ionization potentials and electron affinities. Int. J. Quant. Chem. 11, 149 (1977).

Serafini, A., Poilblanc, R., Labarre, J.-F., Barthelat, J.-Cl.: Non-empirical pseudopotentials (PSIBMOL algorithm) for molecular calculations: the $Rh_2Cl_2(CO)_4$ complex. Theor. Chim. Acta 50, 159 (1978).

Serre, J.P.: Linear representation of finite groups. New York: Springer Verlag 1977, Chapt. 1.

Trogler, W.C., Cowman, C.D., Gray, H.B., Cotton, F.A.: Further studies of the electronic spectra of $Re_2Cl_8^{2-}$ and $Re_2Br_8^{2-}$. Assignment of the weak bands in the 600-350 nm region. Estimation of the dissociation energies of metal-metal quadruple bonds. J. Am. Chem. Soc. 99, 2993 (1977).

Trogler, W.C., Gray, H.B.: Electronic spectra and photochemistry of complexes containing quadruple metal-metal bonds. Acc. Chem. Res. 11, 232 (1978).

Vahrenkamp, H.: What do we know about the metal-metal bond? Angew. Chem. (Int. Ed.) 17, 379 (1978).

Reactivity and Structure

Concepts in Organic Chemistry

Editors: K. Hafner, J.-M. Lehn, C. W. Rees, P. v. Ragué Schleyer, B. M. Trost, R. Zahradnik

This series will not only deal with problems of the reactivity and structure of organic compounds but also consider synthetical-preparative aspects.
Suggestions as to topics will always be welcome.

Volume 1: J. Tsuji
Organic Synthesis
by Means of Transition Metal Complexes
A Systematic Approach
1975. 4 tables. IX, 199 pages
ISBN 3-540-07227-6

Volume 2: K. Fukui
Theory of Orientation and Stereoselection
1975. 72 figures, 2 tables. VII, 134 pages
ISBN 3-540-07426-0

Volume 3: H. Kwart, K. King
d-Orbitals in the Chemistry of Silicon, Phosphorus and Sulfur
1977. 4 figures, 10 tables. VIII, 220 pages
ISBN 3-540-07953-X

Volume 4: W. P. Weber, G. W. Gokel
Phase Transfer Catalysis in Organic Synthesis
1977. 100 tables. XV, 280 pages
ISBN 3-540-08377-4

Volume 5: N. D. Epiotis
Theory of Organic Reactions
1978. 69 figures, 47 tables. XIV, 290 pages
ISBN 3-540-08551-3

Volume 6: M. L. Bender, M. Komiyama
Cyclodextrin Chemistry
1978. 14 figures, 37 tables. X, 96 pages
ISBN 3-540-08577-7

Volume 7: D. I. Davies, M. J. Parrott
Free Radicals in Organic Synthesis
1978. 1 figure. XII, 169 pages
ISBN 3-540-08723-0

Volume 8: C. Birr
Aspects of the Merrifield Peptide Synthesis
1978. 62 figures, 6 tables. VIII, 102 pages
ISBN 3-540-08872-5

Volume 9: J. R. Blackborow, D. Young
Metal Vapour Synthesis in Organometallic Chemistry
1979. 36 figures, 32 tables. XIII, 202 pages
ISBN 3-540-09330-3

Volume 10: J. Tsuji
Organic Synthesis with Palladium Compounds
1980. 9 tables. XII, 207 pages
ISBN 3-540-09767-8

Volume 11:
New Syntheses with Carbon Monoxide
Editor: J. Falbe
1980. 118 figures, 127 tables.
XIV, 465 pages
ISBN 3-540-09674-4

Volume 12: J. Fabian, H. Hartmann
Light Absorption of Organic Colorants
Theoretical Treatment and Empirical Rules
1980. 76 figures, 48 tables. VIII, 245 pages
ISBN 3-540-09914-X

Springer-Verlag
Berlin Heidelberg NewYork

Lecture Notes in Chemistry

Edited by G. Berthier, M. J. S. Dewar, H. Fischer, K. Fukui, H. Hartmann, H. H. Jaffé, J. Jortner, W. Kutzelnigg, K. Ruedenberg, E. Scrocco, W. Zeil

Volume 1: G. H. Wagnière
Introduction to Elementary Molecular Orbital Theory and to Semiempirical Methods
1976. 33 figures. V, 109 pages
ISBN 3-540-07865-7

Volume 2: E. Clementi
Determination of Liquid Water Structure
Coordination Numbers for Ions and Solvation for Biological Molecules
1976. 32 figures, 18 tables. VI, 107 pages
ISBN 3-540-07870-3

Volume 3: S. R. Niketić, K. Rasmussen
The Consistent Force Field
A Documentation
1977. 24 figures, 26 tables. IX, 212 pages
ISBN 3-540-08344-8

Volume 4: A. Graovac, I. Gotman, N. Trinajstić
Topological Approach to the Chemistry of Conjugated Molecules
1977. 34 figures, 2 tables. IX, 123 pages
ISBN 3-540-08431-2

Volume 5: R. Carbó, J. M. Riera
A General SCF Theory
1978. 25 figures, 19 tables. XII, 210 pages
ISBN 3-540-08535-1

Volume 6: I. Hargittai
Sulphone Molecular Structures
Conformation and Geometry from Electron Diffraction and Microwave Spectroscopy; Structural Variations
1978. 40 figures. VII, 175 pages
ISBN 3-540-08654-4

Volume 7
Ion Cyclotron Resonance Spectrometry
Editors: H. Hartmann, K.-P. Wanczek
1978. 66 figures, 32 tables. V, 326 pages
ISBN 3-540-08760-5

Volume 8: E. E. Nikitin, L. Zülicke
Selected Topics of the Theory of Chemical Elementary Processes
1978. 41 figures, 6 tables. IX, 175 pages
ISBN 3-540-08768-0

Volume 9: A. Julg
Crystals as Giant Molecules
1978. 8 figures, 21 tables. VII, 135 pages
ISBN 3-540-08946-2

Volume 10: J. Ulstrup
Charge Transfer Processes in Condensed Media
1979. 32 figures. VII, 419 pages
ISBN 3-540-09520-9

Volume 11: F. A. Gianturco
The Transfer of Molecular Energies by Collision: Recent Quantum Treatments
1979. 17 figures, 6 tables. VIII, 327 pages
ISBN 3-540-09701-5

Volume 12
The Permutation Group in Physics and Chemistry
Editor: J. Hinze
1979. 20 figures, 13 tables. VI, 230 pages
ISBN 3-540-09707-4

Volume 14: E. W. Thulstrup
Aspects of the Linear and Magnetic Circular Dichroism of Planar Organic Molecules
1980. 27 figures, 5 tables. VI, 100 pages
ISBN 3-540-09754-6

Volume 15
Steric Fit in Quantitative Structure-Activity Relations
By A. T. Balaban, A. Chiriac, I. Motoc, Z. Simon
1980. 20 figures, 43 tables. VII, 178 pages
ISBN 3-540-09755-4

Volume 16
P. Cársky, M. Urban
Ab Initio Calculations
Methods and Applications in Chemistry
1980. 23 figures, 54 tables. VI, 247 pages
ISBN 3-540-10005-9

Volume 17
H. G. Hertz
Electrochemistry
A Reformulation of the Basic Principles
1980. 44 figures. X, 254 pages
ISBN 3-540-10008-3

Volume 18
S. G. Christov
Collision Theory and Statistical Theory of Chemical Reactions
1980. 30 figures, 9 tables. XII, 322 pages
ISBN 3-540-10012-1

Volume 19
E. Clementi
Computational Aspects for Large Chemical Systems
1980. Approx. 78 figures, approx. 59 tabels. V, 184 pages
ISBN 3-540-10014-8

Springer-Verlag
Berlin
Heidelberg
New York